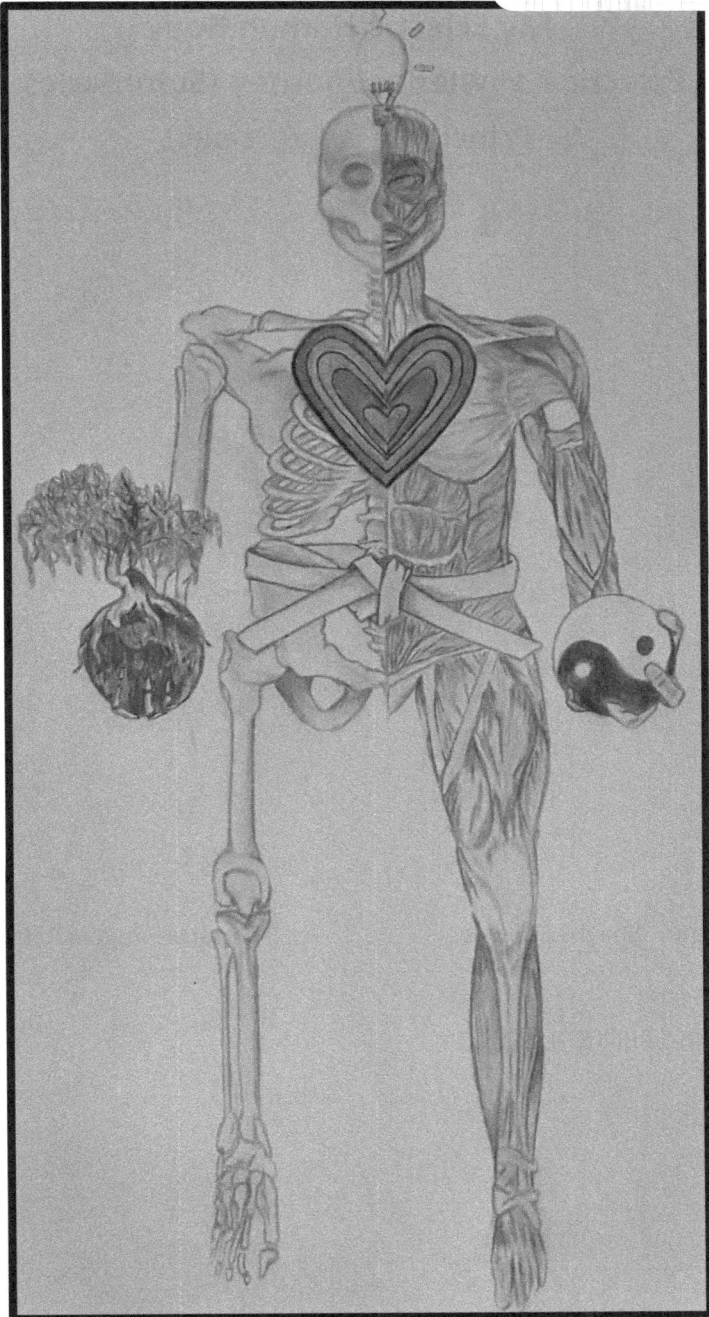

My Eclectic Human Body
Eclectic Knowledge Journey (Summaries, Principles & Exercises)

Author: Shayne T Pattie   Cover Illustrator: Angela Pattie

Editor: Phillip Deguara

2023

In loving memory of Fiona.

Contents

Introduction 7

   Why this Book...7

   About the Author...9

      My Childhood and Teenage Fitness Experiences 9

      My Sporting Injuries 9

      My Various Minor Medical Issues 10

      Family Serious Medical Diagnoses 11

      About the Author Relevant Studies & Qualifications.........11

      Brief Summary of qualifications and other relevant information...13

      Example of Perfecting Health 13

Human Body Basic-Complex.....................................21

   Body Systems 21

      Body Systems Introduction 21

      Circulatory (Cardiovascular) System 23

      Respiratory System..28

      Skeletal System...31

      Muscular System 38

      Nervous System..41

      Endocrine System 48

      Digestive System 52

      Genitourinary and Reproductive System 58

      Integumentary System 62

      Immune System..66

Fascia 72

Summary 74

Tai Chi Practice for Health 76

Qi Gong Understanding 81

Yoga Understanding 95

Holistic Health Information 100

Pilates Understanding 111

Herbalist Understanding 114

Nutrition 131

Nutrition related information and tips...........134

Mental Health and Nutrition 147

Useful Nutrition Information...........................156

Psychoneuroimmunology Understanding 160

Psychology of Body Brain Connection (Stress, Trauma, Enteric Nervous System) 166

Psychology of Human Body Brain Connection Introduction 166

Stress and Trauma Brain and Body changes ...178

Enteric Nervous System (nutrition and mental health) ...186

Examples of Tools (not to be used as prescription) 188

Information to discuss with a medical or allied health professional 203

Psychology of Sensory Systems...........................205

Martial Arts (Philosophies focused)...................212

Focus 212

Itosu Shito Ryu Karate 214

Rhee Tae Kwon Do...215

Wing Chun Kung Fu..215

BJC Muay Thai..216

Zen Do Kai Karate 217

Pressure Point Knowledge..218

Dabbled 219

Self-Studied 220

    Jeet Kune Do..220

    Judo 221

    Ninjutsu 222

    Karate 223

    Philippine Fighting Arts (Arnis/Kali/Escrima) 224

    Aikido..225

    Jujutsu 226

    Hapkido 227

    Krav Maga..227

    Systema 228

Fitness Exercises – Pilates, Gym, Calisthenics and Yoga 230

Pilates Exercises...234

Gym 243

Body Weight, Calisthenics & Gymnastic Rings 252

Yoga...277

Appendix/References 280

Formal References 280

Informal References 297

Previous Influential Professionals......................................304

Introduction

Why this Book

The original idea for this book arose, as I was unhappy with the amount of human body-related knowledge I had been forgetting from both formal and informal learnings. It is aimed to be an informative and practical book that almost anyone can use to improve their generalised knowledge of the human body, and as a reference tool for myself. This is not meant to be a diagnostic tool, instead it is designed to be a helpful starting point, for those interested in the many intersections of the human body. Throughout my journey thus far, I have had many giving and knowledgeable people willing to assist my thirst for knowledge, some of whom I have included in the appendix.

This book will look at the human body in several different ways. The book will begin by discussing the body systems, different cultural paradigms of the human body, the Herbalist understanding of the human body, nutrition and Psychology regarding the brain and body connections and many other ideas. Next the book will look at the philosophies and focuses of various martial arts and self-defence related ideas that I have direct experience with or have self-studied. Finally, this book will finish with a look at the various exercises (including Pilates, Gym Equipment Exercises, Calisthenics, Acrobatic Circus and other) that I have used myself or taught. I have focused on only the exercises, martial arts, philosophies, tools, etc., that I have used either in a personal setting, or in any of my previous and current body related professional settings. Due to the nature of this book, I have included a list of certifications and trainings I have completed, in the appendix, as well as references to various formal and informal sources I have used (including YouTube videos, articles, DVDs, books, etc.). I have tried my best to include all references and courses but there may be some references I have missed by accident. Also due to the nature of the book, I chose not to do any formal referencing as I had difficulty during the writing process, knowing exactly which sources provided which information.

The early sporting achievements discussed next, followed by the various injuries and minor medical conditions, have led to an increased thirst to understand the Human Body from various scientific and cultural paradigms. As I will reiterate throughout the book, the majority of the information in this book is not my original work, rather it is my interpretation of what I have learnt from the many resources found in the appendix. Again, there will be overlap across the different chapters, due in part to the nature of human body related research and teachings. Where possible I have tried to minimise repeating information to the best of my ability.

About the Author

My Childhood and Teenage Fitness Experiences

I was very fortunate to be successful in team sports such as Rugby League, and in individual sports including Athletics from a young age, winning various awards. This perhaps started my interest in the human body via fitness. I was also fortunate to have access to old encyclopedias, and other scientific books from a young age which I enjoyed reading. In my early teens I was fortunate again to be successful in various individual and team sports such as running, triple jump and cricket. Later when I started training in Itosu Shito Ryu Karate, I was fortunate to be very successful in the individual competitions and team related events. My interest towards the human body, slowly increased after the many sporting injuries I experienced and the many ongoing minor medical conditions that began.

My Sporting Injuries

The sporting injuries mentioned here were all relatively minor but have been added as they furthered my thirst to understand the human body. My sporting injuries began at approximately 12y/o when I received my first concussion from an interschool team sport known as European Handball. I was diving for the loose ball and somehow managed to land on the top of my head and bounce off the concrete court we were using. At about 15y/o during school gymnastics, I managed to do an extra half flip in the air leading to a 1.5 backflip. This meant I landed headfirst and once I recovered from the dizziness, I was unable to do any type of backward flip or roll, with this mental barrier still existing today. At 16y/o I slid for a loose soccer ball during an interschool competition and received a knee to my face from two players simultaneously leading to being knocked out. Also, at 16y/o I was attempting to get 'Age Champion' in athletics at school, but after 4 days of events, leading up to the main day of events, my kneecap went to the side, and I was unable to walk. In my lack of wisdom, I attempted to run the 800 event and was unable to walk for the rest of the day. At 18-19y/o I was knocked out on two occasions during Muay Thai training. I have also broken my left foot on at least occasions and right foot at least once during sporting related events. After 5 years of knee issues being dismissed by various bulk billing General Practitioners (GP) and being told the knee pain was only "inflammation", I saw my wife's GP who referred me to a knee specialist, and within 12 months, I had a bilateral Arthroscopy and knee clean-up for both knees, which meant minimal movement and sports for 12 months following.

My Various Minor Medical Issues

As a child I experienced and was diagnosed with Asthma, but as I grew older, this slowly improved into a milder form. The first ongoing minor medical issue as an adult, that was discussed by a General Practitioner was my Geographic Tongue. By itself, it meant many of my preferred foods led to an acid-like reaction on my tongue, which has minimal impact on my health, but is one of many triggers for me to understand the body better. Next, I was diagnosed with MdDS (Mal de Debarquement Syndrome), impacting my balance, noise and light sensitivity, and dizziness. Later Alexithymia (experiencing emotions as physiological responses as the primary markers of my emotion related experiences) was discussed with various GPs. This was followed by my mid-level global muscle related pain. Most recently various food intolerances have been diagnosed leading to an increased list of food I cannot consume (at the time of writing this book, the list included dairy, garlic, onion, white bread and chilli). When combined with Alexithymia, pain from consuming intolerable foods, leads to further physiological pain and negative mental health responses. Despite what I know now about the human body, my posture still is not as good as it can be. I have a slight shoulder pronation, slight anterior pelvic tilt, slight poke neck, and overactive Sternocleidomastoids (SCM). A lot of this is from poor feet support as a young person, but some of this is from ongoing poor posture when working at a desk. As stated, all these events and difficulties, have fuelled my thirst for an eclectic understanding of the Human Body.

Family Serious Medical Diagnoses

I have had various older family members diagnosed with Dementia prior to their death and various older family members die from cancer. I also have various family members who have experienced heart attacks, heart murmurs, severe asthma, arthritis, and other major medical difficulties. However, the biggest motivator to understand some other paradigms of the human body came when my wife's family member was diagnosed with a terminal illness. Unfortunately, the medical model was unable to extend her life beyond 349 days post-surgery, and I was unable to assist in prolonging her quality and quantity of life with the amount of human body knowledge I had at the time, as I had not yet learnt enough about the human body. I have never been in a financial position to be able to study medicine to become a doctor, so have attempted to learn as much about the human body through other studies. I have used the loss of my wife's family member as extra motivation to understand as much as I can about the Human body without being restricted to one specific paradigm or framework.

About the Author Relevant Studies & Qualifications

I completed my Certificates 3, 4 and Master Trainer in fitness and personal training in 2011 through the Australian

Institute of Fitness and later completed my Pilates Instructor course through Pilates Institute of QLD. From 2011 onwards, I worked for several brilliant Personal Trainers and Gym Instructors and eventually started my own Personal Training and Pilates business. I started my martial arts journey in 2002 and continued this journey with varying levels of training intensity until my first child, whereby the martial arts journey became more of a hobby and more academic. I have and continue to study in various areas of health and fitness.

Since 2011, I have also volunteered with several Mental Health organisations. I have also worked in a supporting capacity in several Mental Health organisations editing counselling students work, editing Mental Health courses, and delivering Mental Health First Aid. I completed my Diploma in Counselling in 2014 and my Bachelor of Psychology with group Honours in 2015 and have since worked in psychology for several community organisations as a Mental Health clinician, group facilitator and various roles. I have also worked in private practice as a counselling Psychologist, including intellectual and forensic assessments and have written and facilitated various mental health related workshops. My thirst for knowledge is ongoing and evidenced by the amount of extra professional development courses I have completed and continue to complete in both the fitness and psychology fields. I have done extra study in various fields relating to the human body, with the goal being to have a more comprehensive understanding of the human body for both personal gain and professional knowledge that I can share, when relevant, to the people I help in my work.

Brief Summary of qualifications and other relevant information

Currently I practice as a psychologist with my main role involving counselling. I also complete capacity, intelligence and neurodivergent related, and basic forensic assessments. Due to my thirst for knowledge, I am constantly trying to learn and as a result have completed a lot of extra study in the areas of Trauma, AOD, Nutrition, Neurodivergence and Sensory Related Topics.

I have worked as a Personal Trainer Mentor, Personal Trainer, Group Fitness Instructor, Bootcamp Instructor, Pilates Instructor, Tai Chi for Health Instructor, Mental Health First Aid Facilitator, Black Dog Institute REACH Facilitator and other group program facilitator. I have also previously trained as an acrobatic Circus Student.

At the time of writing this book, I am a Qi Gong Student, Herbalist Student, AMN (Applied Movement Neurology) Academy Student, Martial Arts Student, Calisthenics Student, Yoga Student, 'Yoga for Mental Health' Student and PNI Global (Psychoneuroimmunology) Student.

Example of Perfecting Health

This section will attempt to incorporate many of the areas studied, into a weekly plan. This weekly plan will attempt to use ideas and information from the majority of the book. This paragraph will act as a kind of abstract academically speaking, but also an example plan, that people can adjust to fit their own needs, once they have had a chance to read the book in its entirety and once, they speak with the relevant professionals. I will include my own personal example of what I aim to do in my more successful weeks after each idea.

Daily

Eat enough food for your body. If you are unsure how much you need to eat, you can speak with a qualified dietician to receive more accurate information. One way some people use to measure if they are eating enough without needing to understand calorie counting, macros, etc., is based on how hungry they are. If the person is hungry before bed or wakes the next day very hungry, it is highly likely they are not eating enough food for their body. Stress, Interoception, goals, genetics, medical difficulties, and many other differences may influence how much a person needs to eat and how hungry a person feels. Once the person is eating enough food for their body consistently, they can then start to improve their nutrition. Eating enough nutrients, minerals and healthy fatty acids, and reducing or slowly eliminating refined foods and trans fatty acids on a consistent basis is the end goal. A simple way to eat healthy enough is by eating the rainbow (fruits and vegetables), and ensuring protein from meat or legumes, and carbohydrates from breads or bread alternatives such as potato. Also ensuring enough water and liquids are consumed throughout each day. Temperature, humidity, physical activity, individual difference and many other factors make it difficult to follow a blanket rule, but most people will drink enough water is they are consuming a glass or more of liquid, preferably water.

On an average day I consume approximately 3000 calories. This is a lot of food for most people my age and size, however due to individual difference, my body requires a lot of fuel to maintain a healthy weight. Regarding meals the average day for me is as follows – Breakfast is either 1.5-2 cups of muesli with almond milk, cacao nibs and occasionally includes frozen mixed berries, or I have four eggs on two to three pieces of light rye bread; I have two lunches a day during week days which includes approximately 125grams of meat (often kangaroo or chicken breast), approximately ½ cup of basmati rice, and vegetables such as beetroot, spinach, sprouts, beans, and several others (for simplicity I cook my lunches in bulk, and prepare my vegetables in bulk in containers for the week); for dinner it is often very relaxed depending on what the family want to eat and may include crumbed fish with light rye bread, savoury mince and vegetables with rice, meat with vegetables and pasta, or any other option. I attempt to eat the vegetable rainbow for at least one meal a day, although it is recommended for most people to eat the vegetable rainbow as often as is realistic. When I exercise, I will often have a drink after training that includes ¾ scoop protein powder, magnesium powder, spirulina powder, powdered reds and a flavour of my choosing. Most nights before bed I will have a snack such as a spoon of peanut butter or some bread. Regarding liquid intake throughout the day, I normally ingest at least 2.2litres of water daily, and normally have a cup of loose-leaf herbal tea. My loose-leaf tea involves several herbs and plants mixed in a large container, and I consume a scoop of this mixture. At the time of writing this book this mixture includes Organic Ginger Root, Organic Siberian Ginseng, Organic Turmeric, Organic Hawthorn, Organic Peppermint, Organic Lemon Balm, Organic Marshmallow Root, Organic Rosehip, Organic Motherwort, Organic Australian Oat Straw and Organic Red Clover. After I add a scoop of this tea mixture, I then add a sprinkle of black pepper and mixed herbs to assist in bioavailability of the other ingredients, and a teaspoon of powdered dextrose for flavour enhancing.

Sleep is the next simplest (although not necessarily easiest) to focus on. Sleep hygiene related practices such as journaling, limiting stimulants, and maintaining a routine, are helpful in helping the body recharge, repair and consolidate. Trying to get enough natural light in the morning can become part of a person's sleep hygiene practice as can manipulating when and how the person showers. If a person has more than one shower a day, the 'before bed shower' should be more relaxing and the 'wake up shower' should be more refreshing. The easiest way to do this is with water temperature, where most people find colder showers to be refreshing and warmer showers to be relaxing.

As I have young children, sleep feels like it is never enough. However, I aim to be asleep by 9-10pm, and wake the next morning at 4.30-5.30am. This routine is enough for me most days, however, on days I cannot get to sleep, various sleep hygiene tools such as writing down ideas onto paper, listening to relaxing music or walking around the house for a small amount of time, are practiced, before trying again to fall asleep.

Sensory grounding and mindful breath work can be a useful tool to incorporate before work, school drop off or before stress events to help a person have more tolerance for life's stressors. A more formal practice of sensory grounding may include the use of audio bilateral stimulation via headphones to assist both brain hemispheres and the corpus collosum with inter-brain communication. There are a lot of different mindful breathing exercises but some of the simpler ones include extending the exhale, humming breath, and nostril breathing.

For myself, I attempt to use my auditory sense daily to

help calm the brain via the use of the Audio Bilateral Stimulation, other times just listening to music can be beneficial in the moment to help calm. When this is unrealistic to use due to time, location, etc., I will attempt to use a simplified version of the humming breath or and a simplified version of straw breathing to help calm.

Finally, if time allows, and there is soil or grass that is safe to walk on near the person's residence, use of the Schumann's resonance is helpful. Schumann's research found walking in nature and on grass for approximately 15 minutes daily can assist the body's attunement with the world around and can improve mental and physical health.

Unfortunately, due to work and family commitments, I do not do this as regularly as I have suggested but attempt to do this at least once a week.

At least three times a week

Regular exercise that is safe and that the person enjoys can assist in mental and physical health. Exercise examples can include gym, calisthenics, martial arts, tai chi, qi gong movement practice, yoga practice, Pilates and many more. Mobility stretches and static stretches may form part of a person's exercise or can be practiced as their own 20-minute programs. If possible, adding a body scan to the physical training can improve the individual's awareness of self and have the side effect of better performance and or control.

I attempt to train five times a week, but some weeks this becomes three times due to life events. For me it is normally calisthenic related with exercises borrowed from Pilates and Yoga, and I include the mobility stretches at the start of my training, otherwise I become too lazy to do them.

Occasionally, I train in Muay Thai, but this has become almost monthly at present.

Mindfulness practice assists a person in calming, making better choices, focusing, and has many other benefits. There are many ways to practice this such as the person focusing on their breath, with a basic definition being 'focusing on the one thing, in the present moment, without judgement'. Mindfulness can be done whilst sitting still, moving, balancing, etc., and some people use mindfulness when learning a new skill such as the guitar. It is also very beneficial longer-term as it assists the person in training their brain and developing insight.

As well as the mindful breathing discussed earlier, I try to add my mindful practice to my exercise training. As formal seated and stationary mindfulness practice is difficult for myself, I have found adding it to the training can improve the physical performance and improve better body brain awareness and help the brain calm.

Sound related healing practices can be beneficial and can also be completed in various ways. Traditional sound healing practice is often practiced in yoga, and modern examples include binaural beats and the 'Safe and Sound Protocol'.

I was fortunate enough to be taught about the 'Safe and Sound Protocol' which plays music with specifically adjusted frequencies to assist the brain in longer term calming. I attempt to do this weekly, although the goal is three times a week.

Finally for people with stress stored in their body (most people in the modern world), there are a series of exercises that can be practiced known as TRE. TRE can be used to assist the person in releasing tension that is being stored in the psoas which assists in lessening overall stress levels. Most people find it beneficial to practice before bed, as it may exhaust a person physically and neurologically, although a small number of people have reported that it had a negative impact on their sleep.

I practice an adapted version of TRE and have found this helpful. I aim to do this three times a week, although in practice, it often ends in weekly practice.

At least weekly

One of the most forgotten elements to our health in the modern world is the social needs that almost every person has. Everyone has different levels of social needs and benefit from this need being met. It can include large loud events such as concerts or may include having one or two people over for dinner. Whatever the person's social need is, meeting it will greatly assist in maintaining a healthy self, overall.

As I see people daily for my career, and have a family, my other social needs are minimal, and having a friend visit and train with me is often enough to fulfil my social health.

Monthly if finance and time allows

If a person has time and the finance, massage can be very beneficial. Stress is stored in the body, so by having a remedial type of massage, the physical tension can be released thereby having the positive side effect of less stored stress. Some people also find benefit from Acupuncture, Chiropractic adjustments and other alternatives that help to lessen the physical storage of stress.

At present I do not practice this regularly, but when

finance permits, I have found remedial massage very helpful in lessening stress and improving physical health.

Human Body Basic-Complex

As will be the case throughout this book, I will be focusing on what I have learnt, what I have used, etc. As a psychologist I fully understand a concept known as the 'Dunning-Kruger Effect', whereby the less a person knows about a topic, the more they think they know. With this in mind, this chapter will be focusing on the various understandings and interpretations of the human body from various different paradigms, some of which have a strong overlap and some of which have a correlational overlap only.

Body Systems

Body Systems Introduction

According to the various training and studies I have completed, and majority of the available research, there are between 8 and 11 different body systems, this number varies based on how the systems are defined and thus grouped. The systems I have used the most in my personal and professional life include Circulatory System (also called the Cardiovascular System) including Lymphatic, Respiratory System, Skeletal System, Muscular System, Endocrine System, Digestive System, Nervous System, Genitourinary System including the Reproductive System and Integumentary System. It is noted here that despite lacking any higher-level professional qualifications, I have considered Fascia to be its own unofficial system, as it influences our Muscular and Skeletal Systems, as well as our Nervous System. This idea has recently been supported by professionals more qualified than myself, such as Thomas W. Myers in his book 'Anatomy Trains', where he discusses in detail how and why fascia should be considered its own system, using the phrase the "Fascial Web". This 'body systems' section is focusing on male and female anatomy only and not intersex or other sexual combinations for simplicity purposes.

Circulatory (Cardiovascular) System

The Circulatory System (also called the Cardiovascular System) includes aspects of our Lymphatic system. It relies on a healthy and rapid transport system in the body to – deliver nutrients and oxygen to our cells, carry away wastes, toxins, and carbon dioxide, and allow for the chemical messengers such as hormones to move around the body. When there are fatty deposits in the arteries it can lead to high blood pressure and other serious outcomes, and a blood clot can cut off blood supply to a vital organ or brain.

The average healthy adult heart can be measured as approximately two adult fists and sits close to the centre of the chest. It is protected by our thorax and sits between the two lungs and is known as a muscular pump. The heart is split into halves by the septum and each half contains two chambers separated by a cusped valve. The first chamber is known as the Atrium and leads into the second chamber called the Ventricle. The atria (two atrium) are fed into by large veins. The body's veins hold approximately 70% of the body's total blood supply and contain valves which help the blood travel in the correct direction.

Into the right atria is the inferior vena cava, where deoxygenated blood from our head and upper body is delivered. Blood from the veins fill the right atrium and then during the first part of a heartbeat known as the Atrial Systole, the blood is pumped through into the right ventricle through the Tricuspid valve. During the part of the heartbeat known as the ventricular systole, the right ventricle (which is now full of blood) contracts and forces blood out through an artery called the Pulmonary Artery, which branches off and takes the deoxygenated blood to the lungs, where the lungs will draw out the carbon dioxide and flood it with oxygen. The oxygenated blood from the lungs is then transported to the left side of the heart, entering the left atrium via the Pulmonary Veins. The oxygenated blood is then pumped into the left ventricle during atrial systole via the mitral valve and from this ventricle, it passes into the artery known as the Aorta. The aorta then branches out several times into increasingly tine vessels and the vessel transports oxygenated blood throughout the body.

Due to the heart processes, it is vital that the two parts are kept apart. However, when a hole in the heart forms, the person doesn't receive enough oxygen and becomes exhausted and may appear blue. In the womb, the heart isn't fully formed until the end of pregnancy with some babies being born with the septum still not completely sealed, but in most cases, this is remedied soon after birth either naturally or via surgery.

Arteries have a strong muscular wall and contain elastic fibres that are used to stretch as they receive new blood from a heartbeat. The smooth muscle fibres can relax or contract, increasing or decreasing their inner hollow tube, which the blood flows through. This process explains the potential fluctuation of blood pressure. Lifestyle habits can increase a person's blood pressure due to a decrease in elasticity within the arteries. Habits such as smoking, eating a poor diet, and not getting enough exercise. This makes it harder to lower blood pressure within the artery because the person can't increase the size of the hollow inner tubes. The lack of stretch also means the heart must forcefully push the blood along, which can escalate to heart disease.

Large arteries branch off into smaller versions called arterioles and from there they branch off to form tiny capillaries. Networks of capillaries form the capillary beds that feed tissues and cells from head to toe. Whilst arteries are muscular tubes, capillaries are tiny, featuring a layer of Endothelium made of flat small cells which can become drawn apart when the capillary dilates, which increases permeability. This is vital as it allows for the easy transfer of molecules. It is only in the capillaries where blood can release molecules and take up molecules. Certain capillaries are even more permeable than others, such as in the digestive tract and kidneys. There are also leaky versions of capillaries known as Sinusoids, which occur in the liver where the leakiness functions to allow the liver total access to the contents of the blood.

Veins branch out to form venules in the same way that arteries branch out to form arterioles, but veins and arteries are structurally different, with veins having less muscle in their walls and one-way valves. These valves open to allow blood to flow through towards the heart. If the blood starts flowing back towards our feet, the valves close off to stop this. Veins and venules are not exposed to the same sort of blood pressure as arteries and arterioles and they rely on the blood entering veins from the capillary beds to start pushing the blood along, combined with a suction motion at their other end. During an inhalation, the pressure changes in the chest, drawing blood from the beings into the heart. The contraction of skeletal muscle also helps to get the blood moving through the veins. As the muscles contract, veins which the muscles surround are squeezed, and this forces blood up. This means people need to move around to get the venous blood moving.

Red blood cells, also known as Erythrocytes, account for nearly half our blood's volume. The term Haematrocrit refers to this percentage and you may see it mentioned on blood tests. Males have a slight haematrocrit at 42-52 percent while females have 37-47percent. Red blood cells are made in the red bone marrow and live for around 120 days, before popping at the pressure of tiny vessels becomes too much for them. Packed with haemoglobin (250 million haemoglobin molecules per red blood cell) our red blood cells can carry up to a billion oxygen molecules each. The iron molecules in the haemoglobin bond to the oxygen. This is why people with low iron can be impacted, and how it has a marked impact on how nourished our cells and tissues are by oxygen.

White blood cells (monocytes and later macrophages) form part of our immune system, and unlike red blood cells which have lost their nucleus by the time they enter circulation. White blood cells are fully equipped with their nucleus and are very useful and move around the body. The monocytes also play a secondary role with emotions and research has found that every neuropeptide receptor that could be found in the brain is also on the surface of the human monocyte. Human monocytes have receptors for opiates, PCP, and other peptides such as bombesin. These emotion-affecting peptides, then, appear to control the routing and migration of monocytes, which are very pivotal to the overall health of the human body. White blood cells only account for around 1percent of the total volume of blood and a healthy white blood cell count is between 4800 and 10800 per cubic millimetre of blood.

Platelets, also known as Thrombocytes, are the fragments of what used to be rather large cells called Megakaryocytes. They come into use when we are bleeding with their fragments becoming sticky, forming a plug to stop the blood leaking. Sometimes their ability to form clots can cause problems, such as when the blood doesn't clot well (Haemophilia and other clotting disorders) meaning a risk of someone bleeding to death. Another clotting issue can arise when a clot forms, when and where it shouldn't form, leading to Thrombosis (blocking of a vessel). The position of thrombosis will greatly influence the health outcome. The most common place for thrombosis occurring is the legs, but if they reach the brain, they can cut off the blood supply and cause a stroke.

The Lymphatic System assists the Circulatory System by draining excess fluids and proteins from tissues back into the bloodstream to assist in preventing tissue swelling and protects the Circulatory System from foreign invaders. There is also a newer known body system known as the Glymphatic system which works with Glial cells and connects with the Lymphatic system at the Dura to help clear waste from the brain.

Some common cardiovascular related issues include Hypertension, Arteriosclerosis, Angina, Varicose Veins, and Hypotension.

Respiratory System

The Respiratory System's primary role is to supply our blood with oxygen so it can be delivered to all parts of the body. Normally we breathe in through our nostrils. From here the inhaled air will move deep into the bony cavities behind the nose, all of which feature smaller bones called Turbinates. Their function is to churn the inhaled air so that it communicates with the mucous membranes within the cavity. The sinuses are connected to the nasal cavities by little tubes, and these are prone to congestion when we have an upper respiratory tract infection. The sinuses are lined with mucous membranes which normally help to trap the dust and microbes, but during an infection, all the mucous can trigger annoyance and other emotions.

The nasal cavity communicates with the throat via the Pharynx. The pharynx houses the Epiglottis which functions to close off the windpipe from food and drink. The pharynx divides into the Oesophagus at the back leading down to the stomach and into the larynx at the front for inhaled air. The larynx becomes the trachea, and this is a strong smooth muscle structure supported by horseshoe shaped cartilage coated with more connective tissue.

The inside of the traches is lined with hairy epithelium. Two bronchi emerge from the trachea, and these are also strong, and made of smooth muscle and cartilage, with each one leading to a lung. Within the lung, the bronchus divides and subdivides into increasingly smaller tubes. The smallest tubes are called bronchioles and have no cartilage.

The smallest bronchioles terminate in grape-like structures called alveoli. Alveoli features an elastic membrane, allowing them to fill with air. The alveoli walls are composed of two different types of cells – type 1 alveoli cells (site of gas exchange) and type 2 alveoli cells (septal cells which secrete alveolar fluid). Wandering phagocytes called dust cells patrol the alveoli, and their role is to deal with any dust that makes it deep into the lungs. Alveoli are surrounded by pulmonary capillaries, that have deoxygenated blood run through them from the right ventricle of the heart. The type of alveolar cells assists the alveoli with gas exchange in the lungs. The air that has filled the little alveoli is rich in oxygen and this high concentration of oxygen compared to the low concentration of oxygen in the pulmonary capillaries drives the movement of oxygen molecules out of the air sacs and into the bloodstream. All the oxygenated blood is then transported and used to nourish cells throughout the body. During this the pulmonary capillaries have a high concentration of carbon dioxide, while the lungs should have only a low concentration. This forces the carbon dioxide out of the blood into the alveoli, down the concentration gradient via the diffusion process. We then exhale the carbon dioxide as we breathe out. This exchange occurs over the respiratory membrane comprising the single cell wall of alveoli and the single cell capillary wall.

When diseases such as pneumonia occur, the lungs can become consolidated with fluid which thickens the respiratory membrane, making it much more difficult for the gas exchange to take place. In emphysema, the little alveoli become merged together which decreases the surface area for gas exchange. With milder respiratory illnesses which feature large amounts of mucous, the flow of air into the alveoli is reduced, and cancers and other areas of inflammation can physically block the passage of air. This leads to less oxygen and more carbon dioxide in the body, leading to a more acidic environment in our body.

The lungs themselves are wrapped in two layers of

serous membrane (parietal and visceral pleura) known as the pleural membrane. The space between the two membranes is known as the pleural cavity and this is home to a small amount of lubricating fluid, preventing the two membranes from sticking to each other. Inflammation of the pleural membrane can lead to accumulations of fluid in the cavity known as a Pleural Effusion. When air gets between these two membranes, it's known as Pneumothorax. The lungs are divided into lobes by fissures. The left lung has a superior and inferior lobe only (due to the location of the heart and smaller lung size) whilst the right lung also has a middle lobe.

The two important muscles involved in the process of breathing in and out are the diaphragm and the intercostal muscles. When we inhale, the diaphragm muscle contracts, flattens out and elongates the thoracic cavity; at the same time the intercostal muscles contract and pull the ribs up and out to further expand the thoracic cavity. Due to the change in pressure within the thoracic cavity, the little alveoli in the lungs experience a drop in pressure and air floods into them down a concentration gradient. When we exhale, the diaphragm relaxes and springs back into its original shape. And another set of intercostal muscles draw the ribs back in, which squeezes air up and out of the lungs. The process is known as pulmonary ventilation. For the lungs to perform their job, they need a degree of elasticity, and the bones need a good amount of movement, however, there are illnesses which reduce this elasticity such as Fibrous Lungs Disease, or bone illnesses such as Arthritis, which leads to breathing becoming laboured.

Common issues regarding the respiratory system include Colds and Flus, Sinusitis, Sore Throats, Chest Infections and Hay Fever.

Skeletal System

Skeletal System – There are approximately 206 bones in the adult human body. This section will again focus on the bones and joints that have been and are currently relevant to my professional and personal knowledge. The Skeletal System has many roles – it supports our body weight, allows for movement, produces blood cells, supports immune system, protects and supports organs and stores minerals in the body. In summary bones are a type of living tissue and are responsible for more than just holding us up.

      The human skeleton can be broken into eight sections – the Skull, Spine, Chest, Arms, Hands, Pelvis, Legs and Feet. The main bones that make up the Skull are the: Frontal Bone, Parietal Bones, Temporal Bones, Occipital Bone, Sphenoid Bone and Ethmoid Bone. The Skull has a Tendon known as the Galea that sits over the dome of the Skull. There are 33 bones in the Spinal Column and are often separated into 5 sections – the Cervical, Thoracic, Lumbar, Sacral and Coccyx. There are seven Cervical bones, twelve Thoracic bones, five Lumbar bones, five Sacral bones and the Coccyx made up of four fused bones. There are three main Ligaments that connect along the Spinal Column including the – Ligamentum Flavum, Anterior Longitudinal Ligament and the Posterior Longitudinal Ligament. The Joints along the Spinal Column are referred to as Facet Joints. Joints have often been thought to be passive, but research is now finding that joints are very active in our body and can be strengthened the same way as our muscle – through appropriate training. Our bones are also changeable as will be discussed, but briefly speaking our bones also include the fibrillar part of our bone known as the collagen part, which is often not taught or seen when learning about the skeletal system.

The main bones that make up the Chest are the Thoracic Vertebrae, seven of the twelve pairs of Ribs, and the Sternum. The main Tendon in the chest is called the Pectoralis Tendon, and the seven pairs of ribs connect to the Sternum via their own Costal Cartilage.   The main bones that make up the Arms are the – Humorous, Radius and Ulna bones. The main Ligaments that connect to this area are the – Ulnar-Collateral Ligament, Lateral Collateral Ligament and the Interosseous Membrane. The main Joints that connect to this area are the – Humeroradial Joint, Proximal Radioulnar Joint and Distal Radioulnar Joint.

The main bones that make up the Hands are the – fourteen Phalanges, five Metacarpal Bones and eight Carpal Bones. The main Ligaments that connect to this area are the – Dorsal Intercarpal Ligaments, Palmar Intercarpal Ligaments, Interosseous Intercarpal Ligaments, Pisohamate Ligament and the Pisometacarpal Ligament. The main Joints that connect to this area are the Proximal Interphalangeal Joints and Distal Interphalangeal Joints.

The main bones that make up the Pelvis or Pelvic Girdle are the – Hipbones (Ilium, Ischium and Pubis fused together in adults), Sacrum and Coccyx. The main Ligaments of males and females that connect to this area are the Sacrotuberous, Sacrospinous and Iliolumbar. Females' Pelvis also contain the Broad Ligament and Ligaments of the Ovaries and Uterus. The Joints that connect to this area are the – Sacrococcygeal, Lumbosacral, Pubic Symphysis and Sacroiliac.

The main bones that make up the Legs are the Femur, Tibia, Fibula and Patella. The main Ligaments of the Legs are the Anterior Cruciate Ligament, Posterior Cruciate Ligament, Medial Collateral Ligament and Lateral Collateral Ligament.

The main bones that make up the Feet and Ankle are the – Talus, Navicular, Cuneiform, Calcaneus, fourteen Phalanges and five Metatarsals. The main Ligaments in the Feet and Ankle are the – Plantar Fascia Ligament, Plantar Calcaneonavicular Ligament, Calcaneocuboid Ligament and Lisfranc Ligaments. The main Joints of the feet and ankle are the – Subtalar Joint, Midtarsal Joint, Tarsometatarsal Joint Complex, Metatarsophalangeal Joints and Interphalangeal Joints.

There are at least six different types of bone – the Long Bone, Short Bone, Flat Bone, Irregular Bone, Pneumatic Bone and Sesamoid Bone. The Long Bones have a long thin shape and work as levers to permit movement with the help of muscles such as the Tibia and Femur. Short Bones have a cubed shape and allow movement of areas such as the Wrist, and the Tarsal Bones of the Feet. Flat Bones have a more flattened, broad surface and are made up of a layer of sponge like bone between two thin layers of compact bone and their main purpose is to protect internal organs such as the Brain, Hips and Pelvic organs. Irregular Bones do not fit the above three types and perform various functions in the human body including protecting nerve tissue and providing support for the Pharynx and Trachea. Pneumatic Bones can also be classified as a subset of Irregular bones but contain large air spaces lined by Epithelium. Sesamoid Bones are the bony nodules that are found embedded in the tendons or joint capsules and assist the body through resisting pressure, minimising friction, altering the direction of the pull of the muscle and maintaining local circulation.

The outer layer of the bone is called the periosteum and tendons and ligaments attach to our bones via our fibrous connective tissues (fascia) to facilitate and respond to movement. Under the periosteum is the compact bone, a thin layer, and beneath this is the spongy (cancellous) bone. The spongy holes are full of red bone marrow, the site of blood cell formation. Inside the cavities of our long bone we find yellow bone marrow, which serves as a fat storage facility. Therefore, our bones' secondary roles include making blood cells, storing fats and are a reservoir of calcium and phosphates.

Ossification is the process of bone forming. In the mother's uterus, the tiny bones start out as little cartilaginous forms which eventually undergo ossification. The process of ossification requires calcium, which we originally receive from our mother via the placenta. This calcium is laid down in concentric circles, hardening the structure. At the centre of these circles is the Haversian canal.

There are two ossification processes – intramembranous ossification and endochondral ossification. Intramembranous ossification is where the bone is formed within the fibrous connective tissue membranes by condensing mesenchymal cells. Endochondral ossification is where the bone is formed within hyaline cartilage which is then made into bone. As we mature, our bones grow in length and diameter and undergo remodelling. The main elements of bone formation include osteoblasts which secrete bone ingredients, osteocytes which are mature bone cells that help maintain the daily activity of bone tissue and osteoclasts which breakdown and resorb bone tissue.

When a bone is broken, it heals via a process known as calcification. The first event is the formation of a fracture haematoma (blood clot) which occurs when the blood leaks from vessels ruptured in the break. An inflammation response then occurs where the capillaries grow into the blood clot and white blood cells flood the area and this first stage can last several weeks. The second stage is the formation of the fibrocartilaginous callus. As granulation tissue fills the site, a procallus is formed and fibroblasts and osteogenic cells move in. Collagen fibres are then laid down to connect the two ends of the broken bone. The osteogenic cells transform into chondroblasts which then start to produce fibrocartilage, and this stage can last around three weeks. Next the bony callus forms which involves the replacement of the fibrocartilage callus with bone material. This happens as osteogenic cells transform into osetoblasts, which then start to produce spongy bone trabeculae, which then join portions of living and dead bone. This stage lasts three to four months. The final stage is bone remodelling and involves osteoclasts actively doing their job of breaking down bone cells, so they can be replaced by new compact bone around the periphery of the fracture. Bone healing is important and benefits from its piezoelectrical properties.

Bones are a multidirectional and anisotropic piezoelectric material that exhibits an electrical microenvironment. The electrical signals play a very important role in the process of bone repair, which can effectively promote osteoblast differentiation, migration, and bone regeneration. Piezoelectricity can be defined as the electric charge that accumulates in solid materials such as bone, DNA and various proteins in response to applied mechanical stress. Our bones are seventy percent inorganic hydroxyapatite and 30 percent organic type one collagen, and most of our body is Piezoelectric or has properties of this.

Bone Density is important for everyone and become more important as we age. The factors that work together to ensure the health of our Bone Density include – dietary calcium, vitamin D from the sun and from diet, healthy diet with vitamins and minerals, naturally occurring hormones, and regular weight bearing exercise. Regular weight bearing exercise has been studied and a theory known as 'Wolff's Law' discusses this. According to Wolff's Law naturally health bones will adapt and change to the stress they are subjected to, with heavier loads (within realistic parametres) leading to stronger bones. Wolff's Law was created by German anatomist and surgeon Julius Wolff in the 19th century and has been found to be accurate in most cases but did not discuss how bones can become stronger through changes in our bone geometry.

Tendons act as connectors between muscle and bone and allow for movement, whilst Ligaments connect bone to bone and usually hold structures together for stability. They are made out of connective tissue that have a lot of strong collagen fibres. Joints are complex and are often made up of Cartilage, Synovial Membranes, Ligaments, Tendons, Bursas, Synovial Fluid and Meniscus. The four types of Joints are – Ball and Socket Joints such as shoulder and hip, Hinge Joints such as fingers and knees, Pivot Joints such as the neck and Ellipsoidal Joints such as the Wrist.

The three basic categories of joints are fibrous, cartilaginous, and synovial. Fibrous joints fix together adjacent bones with tight fibres and are almost immovable. Cartilaginous joints have some movement and are made of cartilage. Synovial joints are freely movable joints which feature a joint capsule, with a synovial membrane, articular cartilage, ligaments and in some sites a menisci and bursae. The synovial fluid is a gloopy like substance and lubricates the joint. The more we move the more synovial fluid we make and the less we move the less we make. When a joint is starting to stiffen, it can often be due to less synovial fluid. The cartilage at either end of the articulating bones has no blood supply and relies on synovial fluid to nourish it. Pain, stiffness, pain, and inflammation often are caused by the synovial fluid not reaching the cartilage and the cartilage being damaged as a result. Also, if a muscle around a joint becomes too tight, they can compress the cartilage producing similar symptoms.

Adding to what has been discussed about the skeletal system so far, it is noted that until recently it has been taught that the skeletal system is mostly a continuous compression structure, like a brick wall with the head resting on the 7th cervical vertebra, the thorax resting on the 5th lumbar and so on, down to the feet. However, more recent research into the human body has instead suggested that tension (tensegrity) and compression work together. In this model the bones are seen as spacers, pushing out into the soft tissue and the tone of the tensile myofascial becomes the determinant of the balanced structure, where tension and compression work together to constantly try and achieve a balance.

Common issues in the Skeletal system can include Osteoarthritis, Rheumatoid Arthritis, Gout, and Fibromyalgia.

Muscular System

Muscular System – There are over 600 muscles in the human body. In the fitness world, we first learn about the six muscle groups, and later the core. From there we slowly begin the journey through the complexity of the muscles and become more specific with our own and other people's exercise prescription. The six main muscle groups are the Chest, Back, Arms, Shoulders, Legs and Calves. The core is often included in most exercises as it the powerhouse and should be used with most if not all exercises. This section will look at the muscles that I have learnt and discussed with people. As a result, there will be many muscles not named in this section. This section will also include several Tendons as often the Muscles and Tendons are discussed together with fitness related training or recovery.

Bundles of muscle cells can look stripy and make up the fibres that are bound together to form a muscle. Their stripes are created by the actin and myosin filaments which slide across each other on an ATP fuelled reaction that causes muscle fibres to shorten, contracting the muscle. Each muscle cell is wrapped in a fibrous connective tissue sheath and the bundles are then further sheathed. A whole muscle is also sheathed in a connective tissue known as fascia. This fascia extends at the end of muscles to form the tendons that attach to bones.

When training the chest and shoulders the main muscles discussed are the: Pectoralis Major, The Pectoralis Minor, Sternocleidomastoid, Supraspinatus, Trapezius and the Deltoid Muscle.

When training the Back, the main muscles discussed are the: Sternocleidomastoid, Trapezius, Deltoid, Teres Minor, Teres Major, Latissimus Dorsi, Supraspinatus, Rhomboid Major,

Rhomboid Minor, Quadratus Lumborum, Serratus Anterior, Serratus Posterior and Erector Spinae. The Gluteal muscles can often be placed in this category as well dependent on exercise focus, as can many of the other core muscles.

When training the Arms, the main muscles discussed are the: Deltoid, Pectoralis Major, Pectoralis Minor, Biceps Brachii Short and Long Heads, Subscapularis, Teres Major, Latissimus Dorsi, Flexor Carpi Radialis, Triceps Brachii, Brachioradialis, Flexor Carpi Ulnaris, Extensor Carpi Ulnaris, Extensor Carpi Minimi, Extensor Carpi Radialis, Abductor Pollicis, Extensor Pollicis and Pronator Teres.

When training the legs (upper half) the main muscles discussed are the: Semitendinosus, Biceps Femoris, Semimembranosus, Gluteus Maximus, Gluteus Minimus, Gluteus Medius, Iliopsoas, Psoas Major, Adductor, Abductor, Sartorius, Tensor Fasciae Latae, Rectus Femoris, Vastus Medialis, Vastus Lateralis, Iliotibial Tract, and Piriformis.

When training the calves or lower legs the main muscles discussed are the: Gastrocnemius, Soleus, Tibialis Anterior, Tibialis Posterior, Peroneus Longus, Achilles Tendon, Extensors, and Flexors.

When training the core, the main muscles discussed are the: Rectus Abdominus, Transverse Abdominus, Serratus Anterior, Serratus Posterior, External Obliques, External Intercostal, Internal Obliques, and Erector Spinae. There is a long and unresolved debate about how much of the body should constitute the core. For beginners the focus is normally an unofficial band that could be covered by a small towel width around the stomach and lower back. For more advance fitness people and athletes the core can include as little as the towel area, and as much as the Gluteal muscles to the Latissimus Dorsi from behind and from the Transverse Abdominus to the

Lower Pectoralis Major from in front. This is because more difficult body weight exercises such as the Human flag, require the person to have muscle control over most of their body.

There are at least three categories of muscles in the body – Skeletal Muscle, Smooth Muscle, and Cardiac Muscle. Skeletal Muscles are referred to as Voluntary Muscles as they can be consciously controlled. Skeletal Muscles are connected to our bones via tendons, and within each muscle there are thousands of muscle fibres. The Skeletal Muscles are the muscles responsible for movement. Smooth Muscles are detected in the stomach, intestines, and blood vessels, and are referred to as Involuntary Muscles. The primary purpose of Smooth Muscles is to cause the organs to contract to transport chemicals out of the organs. Cardiac Muscles refer to the muscles in the heart and are important for blood pumping and is also considered Involuntary Muscles.

Nervous System

The Nervous System is divided into the CNS and PNS. The CNS is comprised of the brain and spinal cord and the PNS are all the bundles of nerve fibres that run down through the body from the CNS. The PNS is divided up on a functional basis with the Somatic Nervous System in charge of our special senses and conscious decisions such as movement, the Autonomic Nervous System in charge of the unconscious and visceral activity in our body such as digestion.

Neurons are our nerve cells, and they transmit and receive information. Some send messages to the brain, some from the brain to other areas of the body. There are different types of neurons including: Motor, Sensory, Inter and Pyramidal cells. Each neuron has a cell body with branch like projections called Dendrites and a long tail called an Axon. The Axon conveys a nerve impulse away from the cell body, whilst the dendrites convey a nervous impulse towards it. An axon communicates with another nerve cell, or with a glandular cell or a muscle cell (causing it to secrete or contract). Some axons are covered in a white fatty insulating layer called Myelin which helps to transmit impulses faster.

Nerve impulses travel from neuron to neuron, or from neuron to glad or muscle cell and this is known as an action potential. Sodium and Potassium ions drive the process through a concentration gradient. Because a cell membrane is selectively permeable, there won't be a completely balanced concentration of these ions on either side, which allows for the use of concentration gradient to transmit an impulse. Within neuron cell membranes there are sodium-potassium pumps which pump out three sodium ions for every two potassium ions that it brings in. When a neuron is not doing much, there is a slightly increased positive charge of sodium ions outside the cell. These positive ions form a line on the outside of the cell membrane which then attracts a line of negatively charged ions within the cells against the internal side of the cell membrane. This then creates an electrical charge across the membrane called the Resting Membrane Charge.

When part of the cell membrane becomes excited, that part of the membrane becomes more permeable to sodium ions and the positively charged ions outside the cell suddenly all rush into the cell so that the cell is more positive on the inside compared to the outside. This process is called Depolarisation of the membrane. In response the next section of the membrane becomes excited, and the same thing happens, which excites the next bit and so on in a wave effect. Just as the sodium channels open, so do the potassium channels, but not as quickly. As they allow potassium ions out of the cell, flowing down its concentration gradient, they decrease the number of positive ions inside the cell, restoring the membrane to its resting state. This is known as repolarisation of the cell membrane.

When one of the action potentials reaches the end of a neuron it triggers the release of a neurotransmitter and these cross a gap if it exists between one neuron and the next, and then bind to receptors on the next neuron. These are synapses (where one neuron meets another). Neurotransmitters may have to cross the gap, and this gap is known as the synaptic space. Neurotransmitters are peptides or amino acids and with the CNS they may be either inhibiting or excitatory. Once it binds to a receptor it prevents any other neurotransmitter from binding and asserting its effect. The effect of the bound neurotransmitter ends when the neurotransmitter is either actively taken back into the neuron (reuptake), when it's broken down by enzymes or when it diffuses away passively from the receptor and out of the synaptic cleft.

The process of nerve impulses travelling down one neuron, potentially hopping over a gap, to continue over to the next neuron, is how most of the body's information is moved around. An impulse may be sending sensory information to the brain, or it may originate in the brain and pass all the way down to our big toe, which causes it to wiggle by contracting muscles.

The CNS consists of the brain and the spinal cord, protected by Meninges which are a special membrane. There are three meningeal layers – the Dura, the Arachnoid Mater and the Pia Mater. The Dura is the tough outmost layer. The middle Arachnoid layer is rich in blood vessels and fibres. The innermost layer, the Pia Mater is soft and communicates directly with the brain. Between the middle and inner layer is a layer of Cerebrospinal fluid. This fluid helps to cushion the CNS structures, acting as a shock absorber.

The brain is an organ composed of billions of nerve cells. Some areas of the brain appear grey because of the neuron cell bodies and unmyelinated axons. Some parts look white because of the myelinated axons. Neurons in the brain have the potential to grow many dendrites and this allows for an incredible number of potential connections between neurons. The different parts of the brain include the Cerebrum, Cerebellum, Thalamus, Hypothalamus, Hippocampus, Amygdala, Brain Stem and Spinal Cord.

Cerebrum is the site of our higher functions such as special senses, speech, conscious movement, and awareness. Within the cerebrum is the motor cortex which controls voluntary movement and the sensory cortex which receives information about sensations from the skin muscles and joints. The outer layer of the cerebrum is called the cerebral cortex and contains four lobes – the Frontal lobe (speech, emotions, planning, problem-solving, movement and reasoning), Parietal lobe (movement, orientating ourselves in space, stimuli perception and stimuli recognition), Temporal lobe (speech, memory, hearing, and emotions), and Occipital lobe (visual processing).

The Cerebellum is involved in coordination of movement and balance. The Thalamus receives vast amounts of sensory data and is involved in motor functions, and our emotions. The Hypothalamus is the body's thermostat. It is responsible for the increase in temperature during a fever and plays a role in the function of the autonomic nervous system and the endocrine system and is the conductor of the pituitary gland. The Hippocampus is involved in our learning and memory and converts short-term memories to permanent memories. Amygdala is involved in fear, emotion, and memory. Working with the Thalamus, Hypothalamus and Hippocampus, it forms the Limbic system, also known as the seat of our emotions. Brain Stem extends from the base of the brain and houses the control centres for the respiratory system, heart and is involved in vasomotor control. Spinal Cord continues down from the brain stem, travelling through the spinal column (protected by the vertebra) where it is also protected by the cerebrospinal fluid and meninges. From the spinal cord sprout nerve fibres in bundles known as Sensory ascending and Motor descending tracts that branch off to innervate the whole body

The PNS is comprised of nerves and sensory receptors. Nerves are bundles of nerve fibres wrapped in a connective tissue sheath and are big enough to be seen with the naked eye. They transmit information from the brain and spinal cord to other areas in the body via the wave effect of the action potential. The majority of the PNS includes the 43 different segments of nerves – 12 pairs of Cranial Nerves and 31 pairs of Spinal Nerves. Thirty-one pairs of spinal nerves exit from the spinal cord, and they innervate the skin and muscles on our trunk, arms, and legs.

There are two types of nerves present in the PNS – Motor (transmitting information from the CNS to elsewhere in the body) and Sensory (receive information from sensory receptors and transmit this information back up to the CNS). Since the PNS includes all the parts of the Nervous System outside of the Brain and Spinal Cord, there are many pathways and subsections including the – Somatic Nervous System, Sensory Nervous System, Sensory Receptors, Motor Nervous System, Autonomic Nervous System, Sympathetic Nervous System, Parasympathetic Nervous System, and Enteric Nervous System. Most of these systems interact either directly or through another system.

The Somatic Nervous System can be found throughout the human body. The nerves in this system deliver information from our senses to the brain and involves the voluntary control of body movements via the use of the skeletal muscles. It consists of Afferent (Sensory) and Efferent (Motor) nerves, is responsible for the Reflex Arc, uses Interneurons (connects Spinal Motor and Sensory Neurons) to perform reflexive actions, and interacts with the various other nervous systems.

There are different types and classifications of sensory receptors including Visceroceptors, Proprioceptors, Exteroceptors, Mechanoreceptors, Nociceptors, Photoreceptors, Thermoreceptors and Chemoreceptors. Visceroceptors are found within internal organs and tubes/vessels, and these let the brain know about internal changes such as blood pressure fluctuations. Proprioceptors are found within joint capsules and tendons, and these give the CNS information about where the body is in space. Exteroceptors detect external sensory information and they are also part of the special senses. Mechanoreceptors are sensitive to touch and pressure. There are four types of Mechanoreceptors including Meissner corpuscles, Merkel cells, Ruffini corpuscles and Pacinian corpuscles. Nociceptors are sensitive to damage to tissues such as pain, extreme heat and extreme cold. Photoreceptors in the rods and cones of the retina are sensitive to light. Thermoreceptors are sensitive to temperature changes. Chemoreceptors are sensitive to chemical changes and include receptors for taste and smell as well as visceral receptors that are sensitive to changes in the plasma level of oxygen.

The Autonomic Nervous System is a section of the nervous system that controls our vital functions including breathing, circulation, heart rate, digestion, etc. It is divided into two further divisions the sympathetic (fight/flight) nervous system and the parasympathetic (rest and digest) nervous system.

The Sympathetic Nervous System is involved in the fight/flight response to perceptions of danger. The sympathetic nerve stimulation communicates with the heart to make it beat faster and more forcefully and it raises the blood pressure. It also makes breathing more rapidly and moves blood away from areas like digestion and the reproductive system and diverts it to the skeletal muscles and the brain. It dilates our pupils to allow for more light, so we can see more clearly, and triggers the release of adrenaline. It also liberates glucose and fat stores. It can also cause low appetite and low libido as side effects of the Sympathetic nervous system response. This stimulation is preparing the body for a short, sharp burst of activity, whether it is a genuine physical threat or a perceived stress. This leads to issues when the perceived stress is longer lasting.

The Parasympathetic nervous system is a strand of the nervous system that is for a healthy state of being known as rest and digest. The parasympathetic nerves slow the heart rate and breathing and divert blood flow back to the areas that have previously been abandoned.

Lastly, the Enteric Nervous System contains approximately 500 million neurons and many variations of neurons and is often referred to as the second brain. Changes with our nutrition can have a direct impact on our primary brain's functioning via communication from our gut to our brain via the 'Gut-Brain Axis'. The Enteric Nervous System assists the body's other nervous systems in controlling motor functions, local blood flow, mucosal transport and secretions, modulating immune and endocrine functions and assisting with sensory modulation and control. When the Enteric nervous system is negatively impacted by stress and poor nutrition it can cause various other health related issues in the brain and body.

Common Nervous system related issues include prolonged stress, neurological tension, Muscular tension, poorer mental health including anxiety, mood swings, reduced dopamine and GABA production, reduced serotonin synthesis, reduce melatonin production and subsequent insomnia, impaired absorption of important nutrients, gut inflammation and brain inflammation, inaccurate hunger and craving signalling, dysregulations in the brain, poorer memory, headaches, and migraines.

Endocrine System

The Endocrine System is a series of glands that secrete hormones directly into the blood and it works as a team with the nervous system to manage many different bodily functions. The Neuroendocrine hormones are secreted by neurons into the circulating blood and influence the function of target cells at another location in the body. The Endocrine system has many purposes including – controlling and coordinating the body's metabolism, energy levels, reproduction, growth and development and assisting in the response to injury, stress and mood. There are normally eight important parts of this system including the – Hypothalamus, Pineal Body, Pituitary, Thyroid & Parathyroid, Thymus, Adrenal Gland, Pancreas and Ovary for Women and Testis for Men.

Hormones are mostly peptide molecules that act on specific target cells in other tissues. Steroid hormones are made from cholesterol. By bonding to target cells, a hormone changes the activity of those cells (increasing or inhibiting activity). The hormone responds to negative feedback in the body when there's a raised level of a hormone circulating in the blood, the receptors then feed the information back to the gland which then responds by lowering the output of the hormone.

The Pituitary gland is a pea-sized endocrine gland that is attached to the hypothalamus in the brain, that is made up of two glands – Anterior and Posterior pituitary glands. The pituitary glands are known as the Master Gland as they control or strongly influence many of the body's other glands. Its many jobs including Growth hormone production (which is important throughout life, as it assists in tissue replacement and in the metabolism of fats, proteins, and carbohydrates), Prolactin production (responsible for triggering the production of breastmilk), Thyroid stimulating hormone, Gonadotrophic hormones, Adrenocorticotrophic hormone (triggered by the hypothalamus and stimulate the release of adrenal cortex hormone and is one of the principal stress hormones) and Melanocyte-stimulating hormone (stimulates the skin's melanocytes, which release pigment/melanin).

The hypothalamus produces to hormones and then these migrated down into the pituitary gland, and these are – Oxytocin (helps to eject milk from the breast, is involved in uterine contractions, and is a powerful "bonding" hormone) and Anti-diuretic hormone (increase the reabsorption of water in the nephron, forming a more concentrated urine and conserving water for the body).

The Thyroid gland lies in the front of the neck below the larynx. The gland has "wings" with one on either side of the Isthmus, across the throat. It releases thyroxin and triiodothyronine which stimulates the metabolic rate in cells and stimulates growth. It also releases Calcitonin which reduces the amount of calcium in the blood.

The Parathyroid glands are four very small glands which are sunken into the thyroid gland. They release parathyroid hormone which works alongside Calcitonin to regulate calcium levels in the blood. It promotes calcium absorption in the intestines as well as increasing reabsorption of calcium from the bones. It is the counterbalance to the thyroid. When the parathyroids don't secrete enough of their hormone the lack of calcium in the blood can cause intense, excruciating muscle spasms (tetany) and convulsions.

Adrenal glands sit atop the kidneys. The middle part of the gland (medulla) secretes adrenaline (epinephrine) and noradrenaline (norepinephrine), and these are known as the fight/flight hormones. The outer part of the gland (cortex) secretes steroid hormones called glucocorticoids including cortisone. These hormones are involved in glucose, protein, and fat metabolism, promoting the mobilisation of energy stores so that the body has enough glucose to meet its needs. They also depress the immune system and inflammatory responses to divert energy away to other processes. The cortex also secretes mineralocorticoids such as aldosterone and androgens. Women post-menopause convert the androgens to oestrogen.

An excess of cortex hormones can lead to conditions such as Cushing's disease which presents with hypertension, a moon face, weight gain on the trunk, emaciated limbs, thin skin, and diabetes. A deficiency of the steroid hormones can lead to conditions such as Addison's disease which presents fatigue, diarrhoea, and cardiovascular disease.

The fight/flight response continues, and in our modern life this can become unhelpful. When stress doesn't pass in minutes or seconds the hypothalamus directs a releasing hormone to the pituitary which in turn secretes an adrenocorticotrophic hormone (ACTH). ACTH tells the adrenal cortex to release cortisol and this hormone primes the body to handle the perceived stress, making sure that the stored energy within the body is moved back into the bloodstream as sugar.

Cortisol is an anti-inflammatory hormone, but when the parasympathetic system doesn't start, then it can lead to the body adapting to stress. Symptoms of this include tiredness, weepiness, anxiety, agitation, poor concentration, exhaustion, tired immune system, and mental fatigue.

Gonads in women are called ovaries and their hormones are oestrogen and progesterone. Whereas male gonads are called testes and their hormone is testosterone.

The pancreas is an exocrine gland within the digestive system, but it is also an endocrine gland because it has groups of cells called the Islets of Langerhans which secrete the hormones insulin, glucagon, and somatostatin into the blood. These hormones are involved in the regulation of our blood sugar levels. The thymus gland sits behind the sternum and secretes thymopoietin and thymosin, both are involved in the development of T-lymphocytes. As we age the thymus gland slows, and this leads to physical atrophies. The pineal gland helps with our circadian rhythms and is located in the roof of the third ventricle of the brain in the diencephalon. It receives information about light from our retina, and when we receive less light, it starts to secrete the hormone melatonin, contributing to our sleep patterns. Another hormone secreted by this gland is serotonin.

Common issues in the Endocrine system include Hypothyroidism, Hyperthyroidism and Diabetes.

Digestive System

The Digestive System can be said to start in the mouth with the teeth breaking down food and also includes many other body parts including the lips, mouth, pharynx, epiglottis, oesophagus, cardiac sphincter, stomach, pyloric sphincter, duodenum, bile duct, pancreatic duct, jejunum, ileum, ileocaecal valve, caecum, appendix, ascending colon, transverse colon, descending colon, sigmoid colon, rectum and anus. This system can be aided by some simple behavioural practices including eating when as relaxed as possible, chewing food until almost liquid, ensuring sufficient digestive enzymes, eating a variety of pre and probiotic foods and ensuring sufficient glutamine. Chewing food until almost liquid also allows for the rest of the digestive system to be better prepared. Saliva is released in response to the Sight and Smell of food, after taste, more is released by the Parotid, Submandibular and Sublingual Glands – The Salivary Amylase starts the Enzymatic Breakdown of Starches. Saliva is created under Autonomic control and is stimulated by the Parasympathetic nerves and inhibited by the Sympathetic nerves. Saliva production is slowed during stress states via the Sympathetic Nervous System, which can then lead to digestive difficulties as a side effect.

Swallowing food once chewed, initiates the Autonomic swallowing reflex leading to waves of Peristalsis, which in turn help move the food down the Oesophagus. The Oesophagus is lined with Epithelium and a gooey coating of Mucous to help it withstand the potential abrasive state of food. Food then moves into the Stomach via the Oesophageal Sphincter. When the Oesophageal Sphincter is working well, it stays shut unless food is entering or unless vomit is exiting. When there is a Leak, the Acidic contents can cause reflux and other issues.

Food then enters the Mucous Lined Balloon shaped stomach, and then closed off at the top by the Oesophageal Sphincter and at the bottom by the Pyloric Sphincter.

When food enters the Stomach, the Hormone Gastrin (made by G-Cells) is released which triggers the release of Gastric Juice (composed of highly acidic Hydrochloric Acid, with a pH of 1 and is made by the Parietal Cells). Food is then mixed with the Gastric Juice, Mucous, Gastric Lipase, Water, and Intrinsic Factor. Gastrin also makes the Stomach Churn more to allow for the mixing of the food with the Gastric Juices. Due to the Hydrochloric Acid, the Pepsinogen is converted into Pepsin, which is the Enzyme that starts to breakdown of the Proteins. Pepsinogen is inactive until it is converted into Pepsin, which is a positive, as if the Pepsinogen was always active, it could digest our body's own structural proteins.

Hydrochloric Acid also assists immune function by destroying potentially dangerous microbes. It also triggers the release of Cholecystokinin in the Duodenum, which then triggers the release of Bile and Pancreatic Juice, making the Hydrochloric Acid and important trigger for the cascade of digestive secretions. Some medications that suppress the acidity in the stomach may lead to a weakening of the digestive tract.

After Protein digestion has begun, the food is churned into a more liquid state and is now known as Chyme. After several hours, the Chyme enters the Duodenum of the Small Intestine via the Pyloric Sphincter. This may take from 1-6 hours depending on the level of fat in the meal.

The Small Intestine has 3 parts – the Duodenum, the Jejunum, and the Ileum. The Small Intestine is approximately 6-7 metres long, and squeezes the Chyme by Segmentation, which is where the small segments are closed off at a time to force food along the path. As well as digestion, the Small Intestine is also designed for absorption. The walls of the tube are covered in Tiny Projects called Villi, which increase the surface areas available for the process of absorption. Each Villi contacts Lacteal (lymph duct), and the Lacteal is where the fat is absorbed and sent into the Lymphatic Circulation. Each Villus is covered in its own Microvilli, which further increases the surface area.

The Duodenum secretes Secretin and Cholecystokinin (as above). The Secretin triggers the release of a Bicarbonate-Rich fluid from the Pancreas which then neutralises the acidic Chyme. Secretin also has an Inhibitory effect on the Stomach, telling the Stomach it can slow down. Cholecystokinin promotes Bile Release from the Gallbladder and activates the Pancreas to release Pancreatic Juices, and these two digestive aids enter the Duodenum through the Sphincter of Oddi. Cholecystokinin has other actions. It inhibits hunger (Neuropeptide that communicates with the brain), and it also interacts with Immune receptors across the body. This leads to a dampening of the Immune System.

Pancreatic Juices continue the process of starch, fat, and protein digestion while the Bile gets busy Emulsifying fats, allowing for easier digestion from the Lipase Enzymes. Pancreatic Juices contain Proteases which are an inactive form of a protein enzyme which need to be converted by Enterokinase to prevent self-digestion. They also contain Lipases and Nucleases and Amylases.

Bile is made in the Liver by Hepatocytes and stored in the Gallbladder. Hepatocytes draw Molecules from the blood to make the Bile including Water, Bile Salts (made from Cholesterol), Inorganic Salts and Bile Pigments (derived from the breakdown of Haemoglobin from our old Red Blood Cells.

Bile Pigments give our Urine and Faeces their distinctive colour. If Bile can't enter the Duodenum, we may see evidence of – poor fat absorption (Steatorrhoea/fatty faeces leading to a lighter pigment in the faeces), Yellowing of the skin/Jaundice caused by the building up of Bile Pigments in the Blood and can cause darker Urine.

Chyme passes on from the Duodenum down into the Jejunum and then into the Ileum where the Digestive process is completed. It is this process that allows our body to breakdown and use the healthy vitamins and minerals it needs from good food choices.

Malabsorption can arise for various reasons – lack of Acidity in the Stomach, lack of Pancreatic Juices or Bile entering the Duodenum, or from issues with the lining of the Intestines. This then leads to a lack of nutrients reaching the bloodstream and a lack of nutrients reaching the tissues and cells of the body. This then leads to several symptoms such as general fatigue, through to nutritional deficiencies.

Leaky Gut Syndrome suggests that the damage to the lining of the Intestine allows for large molecules such as Proteins, toxins and allergens (which normally can't cross the Epithelium of the Intestines whole) to squeeze between gaps, and entering the bloodstream triggering an immune response as they are recognised as foreign. It has many other possible causes including low dietary fibre, excess of harmful microbiota, excess alcohol, age, Crohn's disease, Cystic Fibrosis, Rheumatoid Arthritis, Atopic Eczema, HIV, NSAIDS and antibiotics, stress, Small Intestinal Bacterial Overgrowth (SIBO) (measured with specific breath test), and for people with gluten sensitivity, the Gluten Protein may trigger Zonulin (protein that increases the permeability between cells of the wall of the digestive tract). Leaky Gut can cause allergies, auto immune disorders, inflammation, neurological functioning issues, gut bacteria issues, and increase risk of mental health difficulties.

When digestion occurs correctly, the remnants of the absorbed food will now pass down through the Large Intestine (Colon). If there is prolonged stress, the Ileocecal valve and other valves may be impacted, increasing the risk of constipation, haemorrhoids, colitis and IBS. The Colon is divided into the initial Caecum (from which the enigmatic Appendix protrudes), the Ascending Colon, Transverse Colon, Descending Colon, and Sigmoid Colon. A lot of water absorption happens in the Colon and any remaining nutrients will also cross the Epithelium into the bloodstream. However, sometimes digestion doesn't occur properly for many reasons. One of these reasons includes the liver not being able to assist the body in detoxification process well enough as the natural process of the liver turning fat-soluble toxins into water-soluble toxins is interrupted. If a person is having symptoms of nausea, morning headaches, bloodshot eyes, constipation, poorly formed stools, pain in the upper shoulders and under the rib cage or dull skin with boils and infections, it may be related to poor liver functioning. In these instances, various lifestyle changes often need to be made, and some people believe that there are detoxification methods a person can implement to assist the body's natural process. Examples of this includes use of saunas, water therapies, specific enemas, vegetable and specific vegetable juice combinations, skin brushing and hydrotherapy.

If all goes well to this stage, when the Faeces moves down to the Rectum, it stretches the walls and triggers the defecation reflex. Soluble fibres can be Prebiotic, meaning they benefit from the bacteria in your gut. When their balance is changed from – antibiotic use, medications, hormone therapy, illness, or poor diet, we can become vulnerable to a variety of digestive issues and system-wide health issues. Probiotics are helpful short term, but in the longer term, Prebiotic fibres are a very successful method for establishing a good bacterial population.

Common issues that arise regarding the digestive system include Diarrhoea, Constipation, IBS, Indigestion, Excessive Acid, Inflammatory Bowel Conditions, Food Allergies, Food Intolerances and Haemorrhoids. Digestive problems are also strongly correlated with poor mental health – bidirectional. Our digestion requires, or at the very least, benefits from the parasympathetic nervous system being active during digestion. Poor digestion increases risk of insomnia, and can be caused by overuse of medical conditions, antibiotics, low probiotics, and insufficient prebiotic fibre in the diet, and stress related mental illnesses such as PTSD that may activate the sympathetic nervous system (which may slow or temporarily stop digestion).

Genitourinary and Reproductive System

The Reproductive System is part of the Genitourinary System, and is different for men and women, and some people are born intersex where they may have one or parts of both reproductive systems. The female reproductive system is a group of organs including the – Ovaries, Fallopian Tubes, Uterus, Cervix and Vagina, also contains the Clitoris which extends inside and outside the body. Ovulation cycles, menstruation cycles and puberty changes make the female Reproductive System more complex than the more simplified male Reproductive System. The male reproductive system includes the external organs of the Penis, Scrotum and Testicles and the internal organs of the Vas Deferens, Prostate and Urethra.

The female reproductive system is orchestrated by the hormones produced by the ovaries. Oestrogen is an important hormone for the reproductive system and there are several types including Oestradiol (very strong), Oestrone (weaker form) and Oestriol (made by the kidneys from other types of oestrogen and is the weakest. The different types of oestrogen are responsible for Promoting the female secondary sex characteristics (including – breasts, soft skin, female body hair patterns and female fat distribution), Maturation of ovarian follicles, contributing to skin structure, contributing to blood vessel structure and contributing to bone strength. Oestrogens stimulate the cells that have receptors for the hormone, such as breast cells and uterine cells, so that they can increase in number. Oestrogen also stimulates these cells to make more oestrogen receptors, rendering them more sensitive to the hormone.

Each month, women who are of "reproductive age" (which can vary depending on the individual person) have the potential to experience menstruation, which is a cyclical loss of the lining of the uterus. During the first half of a cycle, the ovarian follicles develop and midway through the cycle a mature follicle exits its ovum and ovulation occurs. The ovum is released by the abdominal cavity and most make their way to the safety of a fallopian tube; however, some go astray and can lead to an ectopic pregnancy. At the halfway point in the cycle, luteinizing hormone released by the pituitary, then triggers the empty follicle to mature into a structure called a corpus luteum. A corpus luteum secreted progesterone and this then supports a pregnancy if one occurs by preparing the uterus for the arrival of a fertilised egg. Should fertilisation not occur, then the thickened walls are shed, and this is lost as menstrual blood.

Menstrual and Fertility issues are common, and the cycle progresses in a rough pattern (although, every person is different and may therefore follow their own cyclic pattern). Day 1-4 is the follicular (proliferative) phase, and this is where the follicle stimulates the hormone and a small amount of luteinising hormone trigger the ovary to develop its ovarian follicles. The developing follicles produce oestrogen, and this oestrogen then encourages maturation of the ovum. On Approximately day 14 the blood oestrogen level reaches a certain level; the positive feedback mechanism causes a surge of luteinising hormone. This luteinising hormone surge causes an oestrogen surge which pushes forward the maturation of an egg cell and its subsequent release from the follicle (ovulation). If all goes well, the egg will be taken into the fallopian tube by the wafting of the ciliated epithelium in the tubes. Approximately day 14-28 is when the secretory (luteal) phase, the luteinising hormone continues to stimulate oestrogen production and transforms the empty follicle into a corpus luteum. This secretes progesterone which prepares the lining of the uterus (endometrium) to accept a fertilised egg. The rising levels of progesterone and oestrogen inhibit the hypothalamic-pituitary gonadotrophic system. This inhibition reduces the levels of follicle stimulating and luteinising hormones. If sperm hasn't met the egg, the corpus luteum degrades and the ovarian hormones fall to their lowest point in the cycle. The ischaemic phase occurs where the blood supply to the endometrium is cut, and the lining starts to shed. And finally, on days 1-5 the Menstruation occurs where if an egg was not fertilised two weeks after ovulation, the ischaemic phase will see the start of a period. The first day of the period is day 1 of the cycle.

The male reproductive system often focuses on the prostate. The prostate is a gland, and its secretion is the fluid which is released with sperm upon ejaculation via the urethra. The prostate is located on top of the urethra and is the size of a chestnut. If it becomes enlarged, it can push on the urethra and block the flow of urine from the bladder to the urethra.

The urinary system serves a vital purpose filtering our blood and excreting urea, wastes, and surplus water. The system includes the kidneys, their ureters, the bladder, and its urethra. Each kidney is situated on either side of the spine and has a ureter which runs down to the bladder. The bladder sits in the bottom of the abdominal cavity and when it is full of urine it swells enough to be findable above your pubic bone.

The kidney filters 125ml of blood each minute, and within an hour, all your blood has been filtered through the kidneys. Within a kidney are units called nephrons, which are windy little tubules, and approximately a million exist between both kidneys. Each nephron is contains Glomerulus (small spherical structure that contains glomerular capillaries),

Proximal Tubule (convoluted tube that transfers the filtered substances down to the loop of the Henle), Loop of Henle (U-shaped portion of the tubule which continues from the proximal tubule and draws out more water and salt from the remaining filtrate), Distal tubule (continues from the loop of Henle and is the site of further pH management) and the Collecting duct (near the end of the process).

By the time the filtrate reaches the Collecting duct, all fluids and molecules that can be reused have been passed back into the blood and only toxins and waste remain, and this filtrate is now known as urine. The antidiuretic hormone increases the reabsorption of water in the collecting duct, concentrating the urine. In the absence of antidiuretic hormone, more urine is produced. As levels of this hormone rise, the collecting ducts become more permeable to water and more water passes back into the blood. The collecting ducts channel urine out of the kidney through the renal pelvis via the ureters (the peristalsis moves the urine) and the urine flows down into the bladder for storage. As the bladder fills, its elastic muscular walls stretch. When we urinate, the urine passes through an internal and then an external sphincter down the urethra and out of the body.

Common genitourinary system issues include Urinary Tract Infections, Kidney Stones, four different categories Premenstrual Syndrome (PMS-A, PMS-D, PMS-H and PMS-C), Menopause and Endometriosis.

Integumentary System

The Integumentary System is our body's outer layer and includes the – skin, nails, hair and skin nerves and glands. It has many roles with the primary role being to act as a physical barrier to protect our body from bacteria, infection, injury and sunlight. The other roles the Integumentary System has includes – Heat Regulation, Secretion, Excretion, Sensation, Absorption and Synthesising Vitamin D.

The skin is the barrier between our internal and external environment and is one of our first lines of immune defence. The skin has It has several layers including the Epidermis, Stratum Corneum, Stratum Lucidum, Stratum Granulosum, Stratum Spinosum, Stratum Germinativum and the Dermis.

The Epidermis – the outer layer which is constantly shedding and regrowing. About 60million dead Keratinised cells from the outmost layer of the epidermis are shed each day. Blood nourishes the epidermis to fuel this high rate of renewal and the capillaries come via the dermis. The bottom portion of the epidermis undulates (the papillae) allowing for a greater surface area for the blood supply and for nerve endings. It's the papillae that gives us the skin a swirling appearance and our fingerprints.

The Stratum Corneum is known as the horny layer and is the tough outmost layer of the epidermis. These are the dead cells that a full of keratin. This layer adapts to external forces which is why the stratum corneum is tougher on the feet. The Stratum Lucidum is the clear layer that is beneath the stratum corneum and is made of cells that are almost completely broken down. The Stratum Granulosum is the granular layer and is beneath the stratum lucidum and this is where the cells are broken down, while the nuclei are still intact. Stratum Spinosum is the prickle cell layer made of living cells with membranes intact, with interlocking fibrils extending from the membranes. If the layer is subject to continued pressure, the cells undergo mitosis leading to a growth of new skin in that spot. This leads to calluses. Stratum Germinativum is the germinative layer and is the deepest layer of the Epidermis and is where the cells germinate. As well as epithelial cells, this layer also contains keratinocytes and melanocytes (which pigment the skin).

Finally, we have the Dermis. This is a layer of connective tissue which contains various structures including: blood and lymphatic vessels, hair follicles, sebaceous glands (connect to the hair follicles and the sebum that they secrete, keeps the skin waterproof and prevents it from becoming dry), sweat glands, and in the ears, there are ceruminous glands (which make wax as a defence strategy). The dermis also contains sensory nerve endings such as Merkel's cells, Meissner's corpuscles, and Pacinian corpuscles.

The skin functions as a barrier between our body and the outside world and helps the immune system in this way. The keratinocytes make an anti-viral called Interferon, and the Langerhans cells protect us against microbes that make it through the top layer of the epidermis. The skin is waterproof and stops liquids and germs permeating the tissues beneath and locks moisture and nutrients within us.

The skin also communicates through changes in skin colouring when embarrassed, stressed, shocked or frightened. The skin also plays an important role in thermoregulation and uses the nerve endings in the touch organ (skin).

There are at least two kinds of skin healing – Epidermal healing (superficial injuries) and deep wound healing (penetrates through to the dermis or subcutaneous layer).

Epidermal wound examples include grazes, abrasions, and mild burns. When the skin detects epidermal injuries, the basal cells from the stratum germinativum of the epidermis start to enlarge and move across the wound with cells from each side of the wound moving across until they meet in the middle. This meeting of the cells triggers a response called contact inhibition which stops the basal cells from enlarging and moving further. Meanwhile a hormone called 'epidermal growth factor' is released, triggering the division of basal stem cells to replace the basal cells in the stratum germinativum that originally moved up into the wound. This makes sure there is a solid base beneath the wound.

Type two injuries require a lot more healing because multiple layers are involved. There are four phases to deep wound healing including the Inflammatory, Migratory, Proliferative and Maturation phases. The Inflammatory Phase is when a blood clot forms to seal off the gap/wound. Vasodilation of blood vessels in the vicinity of the wound allows a flood of blood that brings white blood cells to battle infection and mesenchymal cells that will develop into fibroblasts. The Migratory Phase is when the blood clot turns into a scab and epithelial cells move beneath the scab to form a bridge of cells. The fibroblasts originating from those mesenchymal cells delivered during the initial inflammatory response then begin to create scar tissue made up of collagen fibres and glycoproteins. The broken blood vessels now start to regrow, and the wound cavity is filled with a tissue known as granulation tissue. The Proliferative Phase is when a proliferation of epithelial cells underneath the scab to form new skin under the scab. Lots of collagen is still being laid down to provide stability and the blood vessels continue to grow. Finally, the Maturation Phase is when the scab is lost, and the epidermis is now back to the thickness it should be. The collagen fibres become more organised and there are fewer fibroblasts present. Deeper wounds are more prone to infection and prolonged inflammation. Other common skin issues include Acne, Eczema and Psoriasis.

Our Hair has several purposes including protecting our eyes from dirt and water, keeping heat in our body, assisting in the cooling response (sweat on hair), and protecting the skin from other damage such as sun damage. The Hair consists of three parts – Hair Shaft (hair we can see), Hair Follicle (keeps hair in skin) and Hair Bulb (responsible for hair growth).

The Integumentary System has four main Glands – the Sudoriferous Glands, Sebaceous Glands, Ceruminous Glands and Mammary Glands. The Sudoriferous Glands secrete sweat through the skin through the pores and hair follicles; the Sebaceous Glands produce natural body oil; the Ceruminous Glands secrete ear wax; and the Mammary Glands are the glands on the person's chest and in female they produce milk after giving birth.

Common integumentary system issues include skin disease, hair loss, poor scalp and nail infection.

Immune System

The Immune System (including the Lymphatic system again) is immensely complex as it has interactions with most, if not all of our body systems. The body has specific and nonspecific immune responses (cell responses). Nonspecific Immunity cells and structures do various other jobs including immunity. The skin physically protects our internal environment from any substance, particle, or microbe, and has Keratinocytes that produce interferons to protect us from viruses, as well as Langerhans cells that lie deeper in the skin that combat potential microbes or debris that penetrates the outer layer. We have the mucous membranes that act as a barrier with their sticky mucous trapping particles debris and microbes. It becomes a liquid when heated so when our body temperature rises during a fever, the mucous liquefies and then runs freely from the body (such as a runny nose) to expel the trapped microbes from the body. When we suppress the mild to moderate fever with medication, we stop the body's natural response which leaves the mucous sticky and congestion elevates, obviously more serious fevers need to be brought back under control.

Tears are another example of nonspecific immunity, washing away particles that have landed on our eye. Tears a filled with powerful disinfectant Lysozyme. Earwax also contains this, and both physically trap and chemically deal with microbes. Vomiting, defaecating, sneezing and coughing help to get rid of unfriendly particles and microbes. The stomach's Hydrochloric acid destroys microbes.

The white blood cell known as the Macrophage, performs nonspecific phagocytosis on several particles and microbes. They appear as groups at sites where microbes are likely to enter the body such as in the tonsils, in the liver sinusoids (called Kupffer cells) and in our alveoli in the lungs. Their cells change shape so that can physically engulf whatever they are unhappy with, to seal it off completely and break it down. Neutrophils also carry out phagocytosis.

Once the body is away from the infecting microbe, the hypothalamus tweaks the thermostat, and our temperature rises. The body then feels cold and shivery as the blood from the peripheries is drawn to the core to heat up the core. You may look pale because the peripheral vessels have constricted to bring the blood to the core. Following this, you start to feel hot as the body tries to make it inhospitable for the microbe so that the microbe can't replicate as quickly. The higher temperature also tells our immune system to hasten its action. To help the body then cool down so it doesn't do damage, the body will start sweating and you look flushed, to cool the body as quickly as possible. Sweat also helps rid the body of any toxins that may have accumulated. Only when a fever fails to break, especially in young children, that intervention is required.

The immune system also coordinates a more specific/targeted immune response to specific microbes and toxins which involves the white blood cells called Lymphocytes. There are two types known as B-Cells (made and mature in the red bone marrow) and T-Cells (made in red bone marrow but mature in thymus). The neuropeptide receptors on the white blood cells communicate with the other lymphocytes, by interacting through peptides called cytokines, lymphokines, chemokines, and interleukins and their receptors. Some professionals believe that the thymus can be assisted by gently tapping the sternum for approximately twenty repetitions, three times a day. It is believed that the vibrations assist in "awakening" the immune cells.

Each lymphocyte is active against one specific antigen. An antigen is a marker found on a microbe, particle or toxin and the lymphocytes have receptors for the antigens. When a lymphocyte encounters a microbe, toxin or particle which has the antigen that fits its specific receptor, the lymphocyte activates and the body recognises the activation and initiates the fever, fatigue, and other symptoms such as rashes, spots, sneezing, coughing, and vomiting. Once they are active the B-cells and T-cells differentiate into both Effector Cells (go to work against intruder) and Memory Cells (to allow for a quicker response in future exposures of the same antigen).

Effector B-cells make immunoglobulins and antibodies. Antibodies are carefully synthesized protein molecules that have been built to bind specifically to the antigen that triggered the B-cells that made it. These antibodies then bind to the antigen on the microbe or foreign body and this binding renders the invader to be more vulnerable to further immune attack. The effector T-cells are divided into two groups – Killer cells (directly attack antigens) and Helper cells (support the activity of B-cells by secreting Interleukins).

Inherited immunity is the immunity we are born with due to our genetic history and genes. There are two types of acquired immunity – natural and artificial. Naturally acquired immunity occurs when our body responds to an antigen and produces antibodies at the same time producing memory cells which remember that antigen, or through breastmilk, or through placenta. Artificially acquired immunity occurs through vaccinations and immunisations, where the vaccine delivers a measured quantity of dead or deactivated pathogenic particles so that the immune response is triggered without the development of the disease. Immunisation introduces an amount of artificially produced antibodies into the bloodstream, and this infers short term immunity.

In response to infection or damage in a tissue the immune system triggers an inflammatory response. Neutrophils are sent first and are followed by Granulocytes, and both these white blood cells release chemicals such as Prostaglandins and Histamine which ensure the continuation of the inflammatory response. Local capillaries dilate and increase in permeability to allow more blood to an area and to allow the necessary cells needed to repair the damage or to fight the infections. The tissue swells due to this fluid accumulation and the swelling serves a purpose – to help contain the infection, discouraging it from spreading and to protect the area. The fluid in the tissue is turned into a gel by clotting factors further helping to trap the infection. Physiology signs of inflammation include Swelling (fluid leaking into tissues from capillaries), Heat (influx of blood and activity in the area), Redness (influx of blood to the area), Pain (squashing of nerve endings by the swelling tissue) and a Loss of function (swelling and the pain).

The body's immune system has hopefully now kept the infection under control, and now the tissue can begin to repair. Many phagocytic white blood cells enter the area and start a cleaning operation, tidying the debris from the repair job. Neutrophils, then Basophils then Macrophages undertake this job. When the tissue is tidied, the white blood cells leave, and we enter the period of resolution. If resolution isn't achieved, it leads to chronic inflammation such as arthritis.

The Lymphatic system's main role is drainage (channelling away fluid and preventing too much storage of the fluid), and it also cleans this fluid and acts as element of the immune system by neutralizing potential harmful particles and microbes. The Lymphatic system is best at removing larger particles such as proteins and particulate matter from tissue fluid. The lymphatic process starts within the tissue as tiny capillaries and then become increasingly large, passing through lymph nodes, and emptying into the systemic circulation via the subclavian veins (deep vein that is the major venous channel that drains the upper extremities).

The lymphatic vessels are extremely permeable, which allows for the easy uptake of interstitial fluid. The lymphatic capillary networks "snake" into tissue spaces, draining away the interstitial fluid, and this is where cells empty their waste products. Once within a lymphatic vessel that fluid becomes known as lymph. Lymph contains molecules that are too large to enter the blood through blood vessel walls, like proteins, and if there has been cellular repair occurring the lymph might contain pus, dead cells, bacteria, or even cancerous cells that are all taken up from the interstitial fluid. The nodes will help to filter out and deal with this debris so that it won't end in the blood. When fats are broken down in the digestive tract, they enter the lymphatic ducts called lacteal, which then turn the lymph a milk colour.

Lymph moves slowly through the vessel as it relies on our veins upon the squeezing action of muscles. The more we walk and move, the more we promote our lymphatic circulation. If there are issues with lymphatic drainage, it can cause Oedema. Lymphatic vessels have a good capacity for regrowth into an undersupplied area, so if lymph vessels have become damaged or blocked, drainage will be resumed once the new vessels grow.

Lymph nodes are also known as glands, and we often are more aware of these when we are unwell. Hundreds of tiny bean-shaped nodes are scattered around the body. Some areas of the body have clusters of these nodes such as the groin, behind the knee, in armpit and in neck. If they're swollen, we notice them and may feel the need to massage this area, which may spread an infection. Within each lymph node is lymphoid tissue, with reticular fibres, connective tissue, and many lymphocytes. Lymphatic vessels enter a node at one end, and the lymph is filtered through the fibrous lymphatic tissue. The lymphocytes engulf the cell fragments, microbes, cancerous cells, and other foreign bodies. This is known as Phagocytosis, where a white blood cell such as a lymphocyte swallows a potentially harmful foreign body, dissolving it into harmless fragments. If the lymphocytes find a particle they can't break down, it's stored within the node to stop it from spreading in the body.

Common immune system issues include fever, autoimmune diseases, an increase in allergic responses, chronic fatigue syndrome, minor infections increasing in frequency, less tolerance, less ability to recover and poorer health across the entire body.

Fascia

As noted earlier, I consider fascia to be its own system, as it has a large influence on multiple other body systems. Modern yoga instructors also appreciate this and reportedly incorporate the fascia theory as part of the instructor course or recommended readings. A basic definition of fascia is – the thin casing of connective tissue that surrounds and holds every organ, blood vessel, bone, nerve fibre and muscle in place. It can be found throughout the body and scientists as far back as DaVinci were discussing this. There are at least seven (up to twelve) fascial lines in our body connecting everything together as part of the fascial web including the – Superficial Front Line, Superficial Back Line, Lateral Line, Spiral Line, Arm Line, Functional Line and Deep Front Line. Fascia also plays a role in the coordination of movement and communicates with our Autonomic Nervous System and Sympathetic Nervous System. Regarding movement, fascia is covered in nerve endings including Proprioceptors and Interoceptors, and fascia also disperses impacts as a tensegrity structure within the body allowing for constant feedback that allows all the muscles to contract and react appropriately. As well as this, fascia also conveys force between muscles, and is capable of reforming itself in response to common movement patterns in order to strengthen along specific lines (completed by our fibroblasts). This allows for our movement choices to influence our fascia and is an important part of our movement and overall health.

Another interesting idea regarding fascia is that it unofficially has two main different rhythms when it comes to adaptation and change. The first rhythm is the play of tension and compression that communicates around the body as a mechanical vibration, so that it travels at the speed of sound (approximately 1100kph), which is more than three times faster than the nervous system. An example is when a person steps from one room to another where there is an unexpected drop or rise. The nervous system, setting the springs of responsive muscles to the expected level of floor is unprepared for the sharp shock that comes. This is then absorbed instead almost entirely by the fascial system over a fraction of a second, as every nuance of changing mechanical forces is noticed and communicated along the fabric of the fibrous fascial net. The second fascial rhythm is the compensatory speed which can slowly adjust over many years and the body's fascial adjusts and pulls other areas to compensate for injury or weakness.

Recent research into fascia has found there is approximately an 80% correlation between the sites of traditional acupuncture meridian points and lines and fascial plane locations and fascial lines. The idea of meridians will be covered in the Qi Gong section of the book.

Summary

Since the body is a complex organic machine, our various systems all have different roles, but communicate with each other either directly or via other systems. The human body is very complex and our knowledge of it is improving and changing almost daily. Many different cultures have different paradigms they use when investigating and understanding the human body as will be seen in some of the next sections. I also note here that the modern "western" scientific approach is very good at understanding the depth and complexities of one specific area of the human, leading to fantastic specialists. However, the body doesn't work in silos (isolation), and there is still plenty of room for other health approaches to assist in this "western" scientific model as these other approaches are often more eclectic and incorporate multiple body systems, but often do not have the specialist knowledge of any one system.

An example of this interconnectivity can be seen when discussing posture, emotions, and perception. The motor cortex of the brain indirectly communicates with our nervous system (often via the hypothalamus). A drastic change in the motor cortex will have parallel effects on thinking and feeling. A practical example of this is when a person changes their posture from having pronated shoulders to opening the chest and gently drawing the shoulders back. This muscular postural change leads to changes in the motor cortex which leads to changes in conscious thought and emotional perception of self. These changes may include subconsciously changing from feeling downtrodden to feeling more confidence. Another brief example of the body's interconnectivity is the physical outwards expression of emotions. All emotions are expressed through both visceral motor changes and stereotyped somatic motor responses, especially movements of the facial muscles. The social and cultural influences often determine what emotions are encouraged or supressed, however, the physical expression of emotions is similar across most cultures.

Tai Chi Practice for Health

When working in my first psychology job, I was fortunate enough to have the opportunity to complete Teacher training in Dr. Paul Lam's Tai Chi for Health program. The program discussed the benefits of Tai Chi, how to be an Effective Teacher, Safety and Risk assessment, their Stepwise Progressive Teaching Method, Tai Chi Principles and many other tools regarding teaching the Tai Chi program. The lessons learnt in the course can be applied in most areas of professional practicing from helping a client one on one, to facilitating effective workshops. This section will be drawing heavily from the course, as well as other Tai Chi resources.

The most common benefits of using Tai Chi, experienced by practitioners and teachers include – improving health, more effectively managing arthritis, diabetes and other chronic conditions, developing patience, tranquillity and inner balance, being part of something greater than oneself.

Many studies have found, and recent studies are finding that Tai Chi has many physical benefits. One of the focuses during the 'Tai Chi for Arthritis' training was 'fall prevention'. When teaching or practicing the sequence of moves, a person may gain improvement in their self-confidence, awareness of their body and joint and muscular health the person decreases their risk of falling and improves their chances of recovery if they do fall. Increasing awareness of fall prevention not only benefits the individual, but also positively benefits the families of people with arthritis as well as lessening the burden on the public health system.

To get the most out of Tai Chi practice, it is important to have an effective teacher. The first aspect of an effective is

Attitude and this includes – The teacher's passion for Tai Chi as a healing art, The teacher's relationship with the participants, being positive in your own expectations, how you speak and, in the feedback, you give the participants, constructively correcting mistakes, expecting positive outcomes for the participants and using positive energy such as having a genuine smile while teaching.

The second aspect is the skill level of the teacher specifically regarding Tai Chi. Having the ability and knowledge to increase or decrease the difficulty level of what is being taught can drastically influence how participants learn. This can also influence their own level of passion for Tai Chi and may influence them to want to teach Tai Chi themselves. This leads to the Teaching skills of the teacher. This is just as important as the ability to do the moves, and some have argued more important. Teaching the way, the participants best learn is a skill in its own right, and when a teacher can combine visual instruction, with clear verbal instruction and with clearly defined movements to practice, people have the flexibility to learn how they learn – visually, auditorily or kinaesthetically.

Communicating effectively is the next aspect and involves Listening, Speaking Clearly, Recognising Feelings and Applying Tai Chi Principles. Being an effective listener involves being able to understand what the participants are saying before responding, and ensuring you wait for the participant to finish speaking before finishing their sentences or answering their questions. Speaking Clearly involves communicating with our body, feelings and with clear instructions, and being able to adjust how you speak to suit the learner's language and Tai Chi ability. Recognising feelings involves recognising the participants feelings based on their language and body language, but also being aware of your own feelings and how they may influence how you see the situation.

Applying Tai Chi principles involves understanding the foundational principle of Tai Chi namely, Harmony/Balance or better known as Yin and Yang. Once this is understood, then applying the principles of control, smooth and continuous movements, and allow for Tai Chi to have a rhythm. Risk assessment by the teacher is ongoing, as any injury can be a setback for the participants. Risk assessment involves communicating with the participants to assess their ability levels, and being able to adjust exercises to fit the participants' needs. When you find out a participant does have an injury, be careful not to overstep by giving medical advice, ensure any advice suggested is relevant specifically to the Tai Chi practice and within the teacher's own abilities and professional training. This will decrease risk of injury and increase the participants' ability to continue their Tai Chi development and practice.

Lastly Facilitating enjoyment is very important in the role of a teacher. This is best achieved through helping the participants find the intrinsic enjoyment of Tai Chi, guiding them through the initial awkward phases, plateau phases and rough patches and using methods such as the 'Stepwise Progressive Teaching Method'.

Tai Chi is considered an Internal art and has key principles including Outward Movement, Body Structure and Internal and key concepts including Jing, Song, Chen and Huo. Outward Movement refers to the improvement of balance and coordination and internal strength through controlled movements. This controlled movement appears from the outside to have a slow, smooth and continuous flow like water flowing in a river against a gentle resistance.

Body Structure refers to maintaining a supple, upright body that is well aligned and in as straight a vertical line as possible. Various injuries or weaknesses may make this more difficult. One example of this would be when a person bends their knees, their body wants to change from vertical to lessen the pressure on the leg muscles. With practice, a participant can become vertical or close to vertical as their body adjusts to the Tai Chi movement patterns. The second part of Body Structure involves being aware of each step as you transfer weight. The transference of weight consciously from one leg to the other allows for more smooth movements whilst maintaining a vertical posture.

The term 'Internal' in Tai Chi refers to the internal components of the human body, with the mind being a large focus. By integrating the mind into Tai Chi practice, the participant can improve their balance and brain body awareness. This internal awareness also involves controlling the muscle tension to allow for gentle, somewhat relaxed movements and joints, and allows the mind to be focused on the movements being practiced instead of the many distractions modern life has.

Jing can be roughly translated as mental quietness or serenity. When a participant focuses their mind on posture, body awareness, breathing, loosening of joints, and relaxing their mind has less ability to focus on the modern life distractions. For some this can be the hardest part of practicing Tai Chi, once the basic movements are learnt, as the participant may then find themselves focusing on other life tasks such as cooking dinner. By practicing internal arts such as Tai Chi, a person can learn to have some control over their stress mind, to help induce the 'Jing State'.

Song can be roughly translated as relaxation and a sense of loosening and stretching out. For people who can visualise, this can be a useful tool as the participant can mindfully focus on loosening their joints by gently expanding them from within the body. This can then lead to a more relaxed state as the body's tension is released.

Chen roughly translates as sinking. This is used when referring to sinking your Qi to the 'Dan Tian'. The Dan Tian in Tai Chi practice is the central point of everything and is an area that is three finger widths below the belly button. Exhaling can facilitate the sinking of Qi to the Dan Tian which helps the body and mind relax and achieve Jing and Song states. In Tai Chi practice, it is believed that Chen enhances stability, Song and Qi cultivation.

Huo refers to agility or the ability to move nimbly. Agility is developed through regular practice with proper body posture, weight transference, control of movements, loosened joints and strong internal strength. In Tai Chi practice, it is believed that Huo aids in Qi cultivation and improves flexibility.

Tai Chi has many health benefits and is often discussed with a practice known as Qi Gong.

Qi Gong Understanding

The biggest difference between Qi Gong and Tai Chi is the focus. They both focus on the human body physically and can have spiritual aspects, however, Qi Gong normally focuses on the internal body systems more specifically and can be explained as a form of movement and mind using intention and mindfulness to guide the body's Qi and/or to make the Qi work. Qi Gong is often used to assist in a particular area such as the mind, body or spirit. To confuse it further, Tai Chi practitioners often use Qi Gong in their practice, and rarely are the two separated as they are in this book.

In this section I will be drawing heavily from the 'White Tiger Qi Gong' courses I have studied (see appendix), as well as several other Qi Gong resources referenced in the Appendix. It is noted however, that despite there being a strong focus on internal spirituality in Qi Gong, my interpretation of the spiritual aspect of Qi Gong will have a strong physical and mental health focus. It is also noted that Qi Gong is often referenced as a healing art but has been used in martial arts for thousands of years for both healing and destructive purposes and intent. However, as stated this section will be focusing more on the physical and mental healing aspect of Qi Gong. Studies have found at least thirteen physical and mental health benefits of Qi Gong including improving – blood pressure, balance, mild anxiety and depression, quality of sleep, muscle strength, bone density, quality of life, immune system, inflammation, cognitive performance (focus), arthritis and cardio fitness.

Before discussing Qi Gong practice in detail, an in-depth discussion of fascia from the Qi Gong perspective, is important. Fascia is the Connective Tissues involved in the movement system, nervous system, immune system, and digestive system. It is the collagenous/fibrous/connective tissues body wide and provides a tensional force distribution network. It helps compartmentalise and separate body areas and assists with communication of strength, force and tension. It also contains mechanoreceptors that respond to mechanical pressure and formation change. For example, the spinal disks are considered fascia and healthy movement of the spine helps strengthen spine fascia. Fascia lines correlate strongly with the majority of the Meridian lines practiced in Chinese medicine. It is believed that each organ has a fascia layer covering it and that they act as energising chambers for the organs.

Fascia consists of 2 types of fibres (Collagen and Elastic Fibres), Cells called Fibroblasts (more Fibroblasts means stiffer movement), Hyaluronic Acid and Glycosaminoglycan (thickening agent that helps bind to water) and Water (three quarters of fascia is made up of bound water). Hydration of the fascia is important for overall body health. Active and passive movement helps the fascia replace the bound water like a sponge, and gravity squeezes fluid out of the fascia while moving. Heat improves movement in the fascia and body and a well hydrated fascia allows for decreases in inflammation. Acidity in the fascia, thickens fluid and reduces movement and tightens fascia.

Like most of the human body, fascia is very efficient at adapting. This is positive when we have healthy movement behaviours but becomes unhelpful in examples such as poor posture where the fascia will assist in holding the poor posture. Mobility becomes stuck if positions are held for too long due to fascia supporting how we hold the body. Approximately 40% of our muscle load is distributed to other body functions via the fascia and the majority of musculoskeletal injuries relate to fascial damage. Some qi gong practitioners and traditional kung fu practitioners believe that the body's qi is stored in the fascia layers where it works like a cushion to protect the organs.

Following on from the useful fascia information is the underpinnings and innerworkings of Chinese Medicine. There are various areas within Chinese Medical Theory, but it can be summarised through the Yin and Yang Principles, the five-element approach, the three Treasures, and Meridian lines.

The idea of Yin and Yang comes from the Chinese version of creation, namely 'The Dao', loosely meaning 'Created Duality'. It was first written in about the sixth century BCE by Lao Tzu (who was a Daoist philosopher) in his book 'I-Ching'. The main aspects of Yin and Yang include – the idea of opposites always being relative to each other and never completely one or the other and both referencing each other, they consume each other to maintain and create balance, and they inter-transform into each other whereby once one reaches its peak it slowly begins to turn into the other. Yin energy comes up from the earth and Yang energy comes down from the sun and sky. Nature Yin is material – produces form, grows, is matter, contracts, descending and below; where nature Yang is – immaterial, produces energy, generates, is energy, expansion, rising and above. Body Yin is – inferior (below), anterior-medial, front, structure, body and organs, front side that we need to protect; where body Yang is – superior (above), posterior-lateral, back, function, head, skin muscles, back side that has protection and is strong.

Activity Yin is – cold, quiet, wet, soft, inhibitions, slowness, substantial, conversion, storage and preserves; where activity Yang is – heat, restless, dry, hard, excitement, rapidity, non-substantial, transformational and change. Disease Yin is – chronic disease, gradual onset, lingering pathological, cold, sleepiness, pale face, likes hot drinks, likes to curl up; where disease Yang is – acute disease, rapid onset, rapid pathological, heat, restlessness, red face, likes cold drinks, lots of bed covers. The Yin organs include Heart, Lung, Liver, Spleen, Kidneys, Pericardium, Conception (Ren). The function of the Yin organs is to store vital substances such as Qi, Blood, Essence, Body Fluids, and to store only pure refined substances it receives from the Yang organs after transformation of food and air. The Yang organs include Small Intestine, Large Intestine, Gall Bladder, Stomach, Urinary Bladder, San Jiao/Triple warmer, Governing (Du). The function of the Yang organs is to empty and refill, they receive, move transform, digest and excrete. Too much or too little of one, will impact all five elements.

Qi Gong discusses a concept known as the 'Dantian'. This can be defined as the body's energy centres, and there are three Dantians in the human body (Upper, Middle, Lower). The Upper Dantian is associated with the Pineal Gland, is the forehead between the eyebrows (sometimes referred to as the 'Third Eye'), and this is where the body's spirit/Shen is refined and converted into 'Wu Wei' (emptiness). The Middle Dantian is associated with respiration, and the health of the internal organs and the Thymus Gland, is level with the heart and this is where the Qi (vitality) is refined and turned into the Shen. The Lower Dantian is associated with the 'Golden Stove Storage' area of the life force, is located 3 finger widths below the navel and processes the body's Elixir into Jing and is where Vitality begins in the body. These three are also known as the body's 'Three Treasures'.

From the Daoist perspective, the first of the 'Three Treasures' is known as Qi. According to Daoist practice, Qi is the shared matter with the universe that – gives use life and vitality, is our breath, immaterial and material (constantly changing form), nutrient substance that nurtures the body and activates life activity. The Qi Essence is the basic constitution strength, that allows for growth, reproduction and development and is the basis of kidney Qi and is the basis of marrow. The next Qi is known as Pre-Heaven Qi and comes from the parents, promotes growth and the development of the human body. Post-Heaven Qi is known as the nutrient Qi and is inhaled from air through lungs and dispersed through lungs. Yuan Qi is strongly related to Essence Qi, is dynamic, is the basis of kidney Qi and facilitates the transforming of Qi and blood. Gu Qi is the Qi of food and is produced by the spleen. Zong Qi derives from the food Qi with the air and is transformed into Gathering Qi, which then nourishes the heart and lungs and enhances and promotes lung function. Zen Qi is the last stage of transformation of Qi and is transformed using Yuan Qi and circulates the channels and nourishes the organs and originates in the lungs. Ying Qi nourishes the internal organs, and closely relates to blood and flows with the blood in the blood vessels and channels. Wei Qi is a coarser form of Qi and flows in the outer layers of the body. It adjusts the opening and closing of the pores, has its root in the lower burner with the kidney, is nourished by the middle burner in the stomach and spleen and is dispersed by the upper burner in the lungs. Central Qi connects to the Yuan Qi in the spleen and stomach and connects to the post-Heaven Qi derived from food. The many functions of Qi can be summarised as – Transforming, Transporting, Holding, Raising, Protecting and Warming.

From the Daoist perspective, the second of the 'Three

Treasures' is known as Shen. Shen is the appearance of thinking and consciousness as well as the internal Zang Fu essence. It is the heart and mind combined.

Finally, the third of the 'Three Treasures' is known as Jing. Jing is – the constitution and physical energy and is inherited from our parents. It is the material basis for the physical body/infrastructure, nourishes, moistens, and fuels the body, is related to reproduction, is the basal root of the body's energy and is stored in the kidneys and the Dantian.

There is thought to be three main causes of disease – internal, external, and other. Internal causes are related to the emotions when they are prolonged or intense that the imbalance occurs such as anger, sadness, excitement, grief, fear, shock, etc. External causes include wind, cold, damp, dryness, summer heat and fire. Other causes of disease include exercise, diet, constitution, over working, fatigue, sex, trauma, parasites, poisons, and incorrect treatment.

The five elements are another cornerstone of Qi Gong practice. The five elements are Fire, Earth, Metal, Water and Wood. The Fire element is thought to create earth and control metal. In the human body it is the heart and small intestine and generates the spleen and stomach and moves in an upward direction. It is also relevant to our tongue sense, bitter taste, colour red, laughter, joy and love, summer, and the hot environment.

The Earth element – is thought to create metal and control water. In the human body it is the spleen and stomach and nourishes the lung and large intestine, house the Yi and is the centre point of reference for the five elements. It is also referred to as the "provider of all" as it is responsible for creation, transforms food into Qi and balances Yin and Yang. It is also relevant to our Lips and mouth sense, muscle tissue, sweet taste, colour yellow, singing, worry and pensiveness, late summer (harvest) and the humid and damp environment.

The Metal element – is thought to create water and control wood. In the human body is the lungs and large intestine and nourishes kidney and bladder and is contraction. It is referred to as the defensive system, turns Yang Qi into Yin Qi, and is also relevant to our nose sense, skin tissue, pungent taste, colour white, crying, grief, autumn and dry environments.

The Water element – is thought to create wood and control fire. In the human body is the kidney and bladder and nourishes the liver and gallbladder and has downward energy. It is referred to as the development and growth element, moving with fluidity. It is also relevant to our ear sense, bones, salty taste, colours blue and black, groaning, fear, winter, and cold environments.

The Wood element – is thought to create fire, and control earth. In the human body it is the liver and gallbladder and nourishes the heart and small intestine and has expansion energy. It is referred to as the planning and vision element and is also relevant to our eye sense, tendons, sour taste, colour green, shouting, anger, spring, and wind environments.

The five elements and Meridians are intertwined throughout the human body according to ancient Chinese Medicine and according to some Traditional Chinese Medicine. I will quickly discuss the difference between Traditional Chinese Medicine and Ancient Chinese Medicine, before returning to the five elements and meridians. It is thought that traditional Chinese Medicine began between the 1940's and 1950's from physicians who attempted to bring together the many streams of Ancient Chinese Medicine and apply them in the modern academic world. This unfortunately led to a loss of the foundations of Ancient Chinese Medicine despite the best efforts, as eventually it was realised that the models were not able to be used under the more westernised medical and academic model. As a side effect of Traditional Chinese Medicine's attempt, it ignored the non-material elements and underpinnings of Ancient Chinese Medicine. What is now considered Ancient Chinese Medicine began between approximately 200BCE-200CE and was the first systematic approach to healing after shamanic healing and moved closer to nature in its approach. It included connecting with resonance from the natural source energy and aimed at helping the student in developing a deeper understanding of energies and self-healing.

Meridians are the pathways in the body (sometimes referred to as channels) that form the anatomy of Chinese Medicine. Qi and blood circulate to the organs internally and externally via these pathways. Nutrient Qi flows inside the meridians and Defensive Qi runs outside the meridians. Diseases of internal organs will find their way into the corresponding meridians, which then can have a more global negative impact. Blockages to the meridians occur if the network of channels is disrupted, this can affect the local area as well as affecting – the organ relating to the meridian, and the Qi, Jing and Shen flow. A person needs to open and clear meridians, to allow for the energy to flow freely. There are Fourteen important meridians – each of the ten major organs have their own associated meridian, the Pericardium and the San Jiao have their own meridians and the Ren and Du have their own meridians. The meridians penetrate to the Yin and

Yang organs and there are three Yin and Yang meridians on the upper extremities and three on the lower extremities. The meridians run approximately symmetrically and run vertically and bilaterally.

There are four fire meridians in total. The Heart fire meridian connects with the small intestine and lungs and is responsible for – clearing heat, unbinding chest, benefiting voice, relaxing muscles, regulating blood flow, and improving heart rhythm. The Small Intestine fire meridian connects with the gallbladder, bladder, ren, heart and stomach and is responsible for – clearing heat, reducing swelling, benefiting shoulder, arms and neck and lessening pain. The Pericardium fire meridian connects with the Sanjiao and is responsible for – soothing Shen, aiding sleep, and dreams, clearing heat from Qi, improving nutritive and blood levels, harmonising stomach and intestines, unbinding chest, regulating Qi. The Sanjiao/Triple Heater fire meridian connects with all burners and pericardium and is responsible for – clearing heat, improving constipation and ear issues, activating meridian channels and lessening pain.

There are two earth meridians in total. The Spleen earth meridian connects with the oesophagus and spreads across the tongue and is responsible for – regulating the spleen, lessening bleeding, restoring consciousness, harmonising middle Jiao, harmonising the spleen and stomach, resolving dampness, harmonising liver, and kidneys, regulating menstruation, regulating intestines and promoting digestion. The Stomach earth meridian enters the stomach and spleen and is responsible for – lessening pain, harmonising stomach, and spleen, nourishing blood, clearing fire, and reviving Yang.

There are two metal meridians in total. The Lung metal meridian connects with the middle Jiao, diaphragm, lungs, and throat and is responsible for – lessening coughing, wheezing, and transforming phlegm, regulating water passage, benefits throat, clears heat from lungs and moistens lungs. The Large Intestine metal meridian connects with the lung and large intestine and is responsible for – helping to improve constipation and tooth aches, inducing childbirth, expelling wind, clearing heat, reduce swelling, regulating ears, nose and throat, adjusting sweating and plays an eliminating role.

There are two water meridians in total. The Kidney water meridian connects with the kidney, bladder and Ren channel and is responsible for – tonifying the kidney as a base of Yin, Yang and Essence, activating Qi, benefiting lungs, strengthening lumber spine and throat, regulating lower Jiao, warming intestines and harmonising stomach Qi. The Urinary Bladder water meridian connects with the spine and lumber region and is responsible for – working with the Vagus nerve, clearing heat from eyes and warming uterus.

There are two wood meridians in total. The Liver wood meridian curves around the stomach and intersects the Du Channel and is responsible for – moving liver Qi, nourishing liver blood and Yin, regulating menstruation and lessening stiffness in the body. The Gallbladder wood meridian connects with the liver and gallbladder and is responsible for – detoxifying, benefitting the head, improving headaches, shaking and tumour growth, spreading liver Qi and helping sinew and joints.

The Ren (conception), Du (governing) and Dai (girdling) meridians are pathways that do not have acupuncture points. The Ren meridian is part of the primary and extraordinary meridians, runs through the central line of the body, has its own channel and is also part of the primary channels. It is responsible for – tonifying Essence, assisting in conception, warming the spleen, nourishing original Qi, calming spirit and grounding. The Du meridian is part of the primary and extraordinary meridians, runs through the posterior central line of the body and is responsible for – calming the spirit, nourishing the brin, tonifying kidneys and Essence, reducing anxiety and promoting sleep. The Dai meridian connects the upper and lower body, runs along the 'Belt Line' of the body and is responsible for assisting with – issues in the middle aspects of the body including lumbar weakness and muscular weakness and issues in the lower extremities.

As well as helping people improve their overall health, in more recent times, Qi Gong has been used to help people recover from some mental health related difficulties. What we consider mental health is still physical as I will discuss in a later section, but we call it mental health as the brain is very complex, and our understanding of the brain still has a lot of room for improvement. According to Chinese Medicine, what we call high levels of anxiety corresponds with the spleen system and is connected to the element of earth. An imbalanced spleen system leads to various digestive issues and impacts our thoughts such as overthinking. The previously discussed fascia information is also relevant when discussing mental health, as it has been suggested that poor mental health can be found as tension in the fascia. Qi Gong discusses using grounding movements along with the other principles discussed to help a person connect with the earth element directly. These movements focus on twisting and dynamic movements to soften the fascia by removing the emotional tension in the fascia, and the twists can be classified as "Co-activation of functional opposites that create spirals in the fascia compartments".

Qi Gong discussed that being stressed or staying in the Sympathetic Nervous System for too long can impact our body in various ways including – changing skin colour (pale), changing fluids such as urine, increasing our brain waves to the Beta 14-30hz more frequently, emitting Wei Qi and burning up the Yang Qi supply, weaking the immune system, depleting Jing, shortening life span, decreasing libido, increasing the risk of high blood sugar levels and type 2 diabetes and leading to an addiction of the negative emotional state.

It is believed that our emotions are chemical consequences that form feedback of our past experiences, that can impact the now and increase our risk of future stress. This can be known as an 'Emotion Loop' and involves – our thoughts driving our feelings, our feelings then driving our thoughts, then the loop hardwires our brain into the same patterns which conditions our body to the past. Then those thoughts cause a biochemical reaction in the brain that release chemical signals, the signals can then make the body feels exactly the way the person was just thinking. This then causes you to generate more thoughts that make you feel the same way you were just thinking. The simplest way to stop the loop is through – focused intention with an elevated emotion we want to replace the previous emotion with, which can then lead to transformation.

The last area of Qi Gong that I found interesting and somewhat relevant to assisting our overall health is the idea of healing sounds. The idea was formally created in 420-589CE by Tao Hongjing, following this, in 1386-1644 various body movements were formally introduced by Hu Wenhuan and Gao Lian. The exact pronunciation of the healing sounds is less important and can vary depending on dialect origins, but it was reported that the common thread of the healing sounds based on the five elements of Chinese Medicine and associated body organs. The healing sound for the heart was suggested to be 'Haa', the lung healing sound was 'Ssss', liver healing sound was 'Shoo', spleen healing sound was 'Whoo', Kidneys healing sound was 'Chway' and the tiple warmer healing sound was 'Shee'. The idea of healing sounds is also explored in yogic practice.

One area that is not specifically associated or taught in Qi Gong practice is known as the Kuji-Kiri or nine syllables/seals. I have included it in this section as it involves Qi cultivation and Chakra awareness and activation. It traces its historic origins to Japanese ninja and samurai practice and each of the nine seals have their own mantras the practitioner would use to focus or channel their mind/spirit/energy. These nine seals include Rin, Pyo, Toh, Sha, Kai, Jin, Restu, Zai and Zen seals. These seals were often used to assist in personal cultivation of Qi as well as the activation of personal chakras. However, there was still a lot of mythology and mystery regarding their use, and many thought they were used as magic spells and as a form of weaponry.

The Rin Kuji is thought to channel power throughout the individual's body to help create focus and generate energy. It is thought to be connected to the fire element in Qi practice and the Root Chakra. The Pyo Kuji is thought to be the energy direction Kuji and is associated with the Sacral Chakra. The Toh Kujiis known as the Harmony Kuji where the person is in harmony with themselves and their environment and is associated with the Solar Plexus Chakra. The Sha Kuji is thought to be the healing Kuji through increasing the individual's own vibration and is associated with the Heart Chakra. The Kai Kuji is thought to be the Premonition of danger or the Intuition Kuji. The Jin Kuji is thought to be Awareness Kuji. The focus of this Kuji is not on the potential for danger like the Kai Kuji, but instead is the awareness of other people's thoughts and possibly emotions. The Jin Kuji is associated with the Third Eye Chakra. The Restu Kuji is thought to be about the perceived control of time and space and is also associated with the Third Eye Chakra. The Zai Kuji is thought to be concerned with Elemental Control. Finally, the Zen Kuji is thought to be Absolute Enlightenment whereby the individual has overcome each of the internal challenges of the other eight, has no hand signals and must be uncovered individually. One possible healing art that utilises Kuji practices is known as Reiki, although this has not been included in this book.

This overlap of Qi and Chakras is interesting as is the history. The Kuji-Kiri's history can be found through the Hindu and later Buddhist spiritual practices, which was later adopted and adapted further for use in the Japanese military and martial practices of the Samurai and Ninjutsu. The Qi element of the practice finds its origins in the Buddhist spiritual practices and the Chakra focuses originate in the Hindu spiritual practices. This overlap is very useful as it demonstrates the overlap of ideologies and health practices across time and cultures and highlights how interconnected or eclectic our human body should be for overall health.

Yoga Understanding

Yoga originates from traditional Indian medicine practices and can be defined as a system of practices that are used to balance the mind and body through exercise, meditation, breath work and developing insight into the person's own emotions and health. Many people also practice yoga as a form of individual spiritual practice. There are many different types of Yoga, with the most common ones including Hatha Yoga (Sun and Moon or Will or Force), Yin Yoga (Restorative), Vinyasa Yoga (Sequence Yoga), Power Yoga (Muscular Endurance), Bikram Yoga (Hot), Ashtanga Yoga (Energetic/Physical), Kundalini Yoga (Spiritual, Sound Healing) and Yoga Nidra (Sleep). My focus as a yoga student has always been relating to the physical exercise component, as such I will be focusing on the physical and mental benefits of yoga, with minimal focus on the spiritual. The easiest way I have found to explain yoga's physical benefits when compared to other exercises and sports is – "improving one's strength through body control and stretch". This definition exemplifies the physical benefits of yoga but fails to encapsulate the two other main reasons why some people are drawn towards yoga, these being breath and spiritual healing/insight development.

Yoga improves our physical health by improving at least six of our body systems including the – Skeletal System, Muscular system, Nervous System, Cardiovascular, Respiratory, Digestive System and brain health.

Yoga improves the Skeletal system by stimulating the release of synovial fluid and improving bone density. Yoga improves the Muscular system by improving the strength and endurance of our supporting/smaller muscles around our joints, as well as placing our larger muscles under healthy levels of strain. Yoga improves the Nervous system by helping to teach a person about their parasympathetic state, and providing the practitioner tools they can use when in a sympathetic state. It has also been found to increase the endorphin related hormones, decrease cortisol and stimulate the Vagus nerve. Yoga improves the Cardiovascular system by assisting in lowering blood pressure and better utilise oxygen through breath training. Yoga improves the Respiratory system by assisting in increasing lung capacity, reducing breaths per minute and improve the flow of the lymphatic system which assists in removing toxins. Yoga improves the Digestive system by decreasing cortisol, decreasing unhealthy cholesterol, boosting healthy cholesterol and lessening the severity of IBS and constipation. Yogic breathing practice has also been demonstrated to assist the brain directly through increasing levels of Gamma-Aminobutyric Acid (GABA) in the brain, developing and strengthening the Interoception regarding emotions and allowing a person to slow down which may positively influence mental health. As a result of all these benefits, it makes sense that our immune system also benefits, and some studies have suggested that the combination of exercise, breath work and development of Interoception, all contribute to this improvement.

As mentioned, an important part of yoga practice is the breath work. The two most common breath practices in yoga are Kapalbhati and Ujjayi Pranayama. Kapalabhati can be translated as 'Shining Skull' and is a method used to warm up via breath. It involves controlling the breath by sharply exhaling while pumping you stomach in and out. The inhalation is passive, while the exhalation is forceful and sharp. This is theorised help the lungs clear any waste from the air passageways. Ujjayi Pranayama is often referred to as an 'Ocean Breath' and is used to calm the brain and create internal heat. As well as learning various ways of breathing, the person often learns various positions or sequences to help them reach their emotional, spiritual or physical goals. There are many examples of yogic breath work, and some of the other examples will be discussed in the 'Psychology of Body Brain Connections' section.

Another important part of most yoga practice involves a basic understanding of what is known as Chakras. Chakras can be defined as wheels or disks that allow the main energetic channels of the body to flow through and from. There are more than 100 Chakras, and with the seven main Chakras/Energy Centres located at approximately the Root, Sacral, Solar Plexus, Heart, Throat, Brow and Crown. Some yoga instructors believe that the seven chakras allow for the 72000 energetic channels called Nadis to carry vital life force energy, often referred to as Prana throughout the body. There are three major Nadis – Ida, Pingala and Sushumna, and they run up and down the spine in a DNA-like helix, and it is theorised that the Chakras form where the three Nadis interact. The biggest different between the Nadis and the Meridian lines, and the Nadis and our nerve systems is that the Nadis are not physical, but more energetic or spiritual channels. This makes it easier to understand spiritually, but harder for it to be understood from the western scientific method.

Another concept of yoga that is useful to understand for physical and mental health reasons is known as the four Bandhas. According to yogic teachings, the four Bandhas can be used to engage, employ and control the body's energy system and to direct this energy to the parts of the body the person wants to and assist the brain centres, Nadis and Chakras. The four main Bandhas include the Root Lock/Mula-bandha, Throat Lock/Jalandhara-bandha, Lifting of the Diaphragm Lock/Uddiyana-bandha and accessing all three locks simultaneously/Maha-bandha, and the two minor Bandhas include the Hand Lock/Hasta-bandha and Foot Lock/Pada-bandha.

The Mula-bandha is activated by exhaling while engaging the pelvic floor and drawing it upwards towards the navel. It is used to activate deep core strength and strengthen the entire pelvic floor area including the muscles that support the pelvic organs. The Jalandhara-bandha is activated by extending the neck while lifting the heart, then dropping the chin to the chest, while the tongue presses into the roof of the mouth. It is used to tone the muscles of the neck and assist in controlling the stream of energy through the nerves and energy channels of the neck. The Uddiyana-bandha is activated while in seated the position. The person will have their feet straight in front of them at hips width. From there, the person will slowly angle the body forward from the waist while keeping a soft joint (small bend) in the knees. The palm of the hands will be placed on to the knees. Next the person takes a deep inhale while pushing the stomach forward, and then use a forceful exhale to empty the lungs. The abdominal muscles are tightened (drawn in towards spine). The position or pose is held for until the next inhalation and process is repeated. This bandha is used to strengthen and tone the abdominal muscles, while practicing meditative, controlled breathing to assist in energizing the body.

The Maha-bandha is achieved through the use of the three major bandhas together (in succession). The Jalandhara-bandha is completed, followed by the Uddiyana-bandha and finally the Mula-bandha. The person releases this series of locks in reverse when they are ready to allow for controlled flow of the body's energy. It is believed that this practice assists in strengthening the autonomic nervous system and pelvic region, assists in anxiety and purging related difficulties, supports the intestinal functioning, assists the body's immune system, assists in thyroid regulation, strengthens internal organs, promotes core strength and energises the body.

Some schools of yoga also discuss the idea of healing sounds in relation to the body's own energy fields (also known as resonance) and chakras. Yogic practitioners who use healing sounds believe that it can directly influence the body's own energy fields and assist in realigning the body's chakras.

Holistic Health Information

This section will focus on the Holistic Health information I found the most interesting and relevant to my job as a psychologist and in my personal fitness and human body goals. This includes information about how light affects Melatonin production, movement ideas, brain information, allergy information and others. The information in this section will be from the AMN Academy Holistic Health Course, as well as other resources.

Our brain operates with five different frequencies, all of which have their own associated states. The Delta frequency is from 0.5-4 hertz and is associated with – deep dreamless sleep, automatic self-healing, immune system function and collective consciousness. Theta frequency is from 4-6.5 hertz and is associated with – deep mediation, light sleep, REM sleep, the dream-like state between sleep and awake, intuition, memory and vivid visual imagery. Alpha frequency is from 7-12.5 hertz and is associated with – calmness, open focus, relaxed thinking, reflective thinking, creative thinking, visualization and effortless learning. Beta frequency is from 13-38 hertz and is associated with – normal waking states, daily activities, close focus, alert/working state, five physical senses awareness. Lastly, the Gamma frequency is 35-40 hertz or greater and is associated with – peak focus and concentration and problem solving. It has been suggested that there are ages when the brain frequencies develop. Delta frequency is developed from 0-2 years of age, Theta from 2-6 years of age, Alpha from 6-12 years of age and Beta from 12 years of age and older.

Sleep performs an important role in our brain health where it serves as a natural detoxification process. During quality sleep many changes happen that include – our brain's cells shrinking, allowing the cell space to expand by more than 60%, flushing waste into the bloodstream via the glymphatic system into the liver to detoxify, washing out Beta-amyloid plaque twice as fast compared to sleep state and increasing flow of the cerebrospinal fluid. Most people follow a monophasic sleep pattern where they sleep for one period and wake for one period. This is useful for most people, however, there is a small percentage of people that find monophasic sleep, less helpful. For this small percentage of people, polyphasic sleep may be helpful. There are variations, however, as long as the person has multiple periods of sleep and wake in a 24-hour cycle, it can be classified as polyphasic sleep.

It is believed that 'Thought' is formless energy until attention is added. This means that our emotions are the sensations we experience from subtle physiological responses which occur subconsciously before we then consciously have the experience. Whilst we can't control this, we can influence this process through various types of body training that involving developing our internal body awareness through various practices such as mindfulness, meditation, yoga, etc. The idea that emotions and thoughts are energy can be discussed using the yogic framework including the seven yogic Chakras.

Using a neurophysiological lens, the seven yogic Chakras have been found to have strong anatomical correlations with the 'Western' medical understanding of the human body, with the Chakras being situated over the locations autonomic Ganglia (clusters of nerve cell bodies projecting from the spinal cord to other nerve cells, which in turn project to their target organs) of the Spinal Cord. The seven Chakras/Energy Centres are Root, Sacral, Solar Plexus, Heart, Throat, Brow and Crown. The Root Chakra is associated with the colour red and is associated with various physiology including – Large Intestine, Rectum and Adrenals. The Sacral Chakra is associated with the colour Orange and is associated with various physiology including – Prostate, Uterus, Testes and Ovaries. The Solar Plexus Chakra is associated with the colour Yellow and is associated with various physiology including – Stomach, Spleen, Pancreas, Liver, Gallbladder and Small Intestine. The Heart Chakra is associated with the colour Green and is associated with various physiology including – Heart, Lung, Thymus and Immunity. The Throat Chakra is associated with the colour Blue and is associated with various physiology including – Lungs (volume of the voice), Speech Organs and Thyroid. The Brow Chakra is associated with the colour Purple and is associated with various physiology including – Hindbrain, Midbrain, Eyes, Pituitary and Pineal Glands. The Crown Chakra is associated with the colour White/Purplish White and is associated with various physiology including – Neocortex and Pineal Gland. Chakras have also been found to correlate with the static electric fields along the spine, which may have a correlation to the spinal autonomic ganglia. It is believed that if a particular Chakra or static electric field is out of resonance with the others, it can be considered "out of phase".

An interesting idea that came from the AMN course was the idea of locking one's joints. For most of my life I had been instructed and trained to always have a slight bend in my joints ('Soft Joints') when performing exercise. However, recent research and understanding suggests that during most calisthenic type movements, it is detrimental to have a Soft

Joints as it impacts the tendon and muscle health. So, by keeping locked joints during calisthenic holds actually strengthens the arm and tendons, although the person has to be careful of hyperextension.

A tight muscle is a muscle that is over facilitated, and a weak muscle is one that is inhibited in its facilitation. By looking at muscles in this way, we can change how we see muscle tightness and change the treatment accordingly, to include mobility, various forms of stretching and strengthening of the weaker muscles. Our Skeletal Muscles contain Muscle Spindles which provides moment to moment information on length and change of length in muscle tissues and play a key role in proprioceptive flexibility. This information is then sent to interneurons in the spinal cord and then the brain stem and cerebellum and basal ganglia which in turn form part of the unconscious proprioception. Another important proprioceptor that influences this proprioceptive flexibility is the Golgi Tendon Organ (GTO). The GTO live at the Muscle Tendonous Junction which is responsive to tension in the muscle which can then reduce type 1a muscle and turn down the volume on the Muscle Spindle to increase flexibility. The term flexibility can be defined at strength at end ranges. Injury can have a negative impact on flexibility, and it is often the pain that has the largest negative impact. The pain we are experiencing can be discussed as part of the Nociception response.

Nociception refers to the Central Nervous System and the Peripheral Nervous System's processing of noxious stimuli, such as tissue injury and temperature extremes, which both activate the Nociceptors and their pathways. Often this is the body's responses to keep us safe from further damage, however, the Nociceptors can be triggered by nerve misfiring and other unhelpful reasons. This leads to a subjective experience of pain a person feels. Improving our Proprioception can assist the body in overriding some of the unhelpful effects of the Nociceptors, as can improving our overall brain body connection.

One simple exercise that can improve muscle control and strength output, that was adapted by the AMN academy is known as Saccade Stacks. Originally researched and used in psychology to help people manage and unlock mental health difficulties, the fitness version is known as Saccadic Eye Movements. AMN academy's research has found the Saccade Stacks can be used to train the person's brain and can increase the motor output of the body and can be completed without a second person being needed, through the use of the person's own thumbs.

As humans, we often forget that we are still animals and that we are just as affected by the sun and planet as any other species of animal on earth. Natural light stimulates the serotonin in our body which, in turn assists with positive/happy moods. Artificial light is rarely full spectrum and light, and even when it is, it can impact our circadian rhythm which in turn can be a risk factor for cancer, depression, earlier death, PMS, PMDD, menopause, andropause, dementia, insomnia, mood disorders type 2 diabetes, and other. Recent western scientific research provides the knowledge to explain how the sun and earth interact with our bodies. The sun has an eleven-year cycle which impacts the level of UV light we receive, and the cycles are correlated with various physical and mental health difficulties.

UV light is a type of electromagnetic radiation, and we normally receive this from the sun. UV light is filtered out from the earth by the ionosphere, and stratosphere. The three types of UV light are UVC, UVB and UVA. UVC light doesn't reach earth and approximately half of the UVB light from the sun reaches earth. All of the UVA light reaches the earth. Visible (VIS) light represents forty-two percent of the total sunlight reaching the earth's surface, Infrared (IR) light represents forty-eight percent of the total light from the sun reaching the earth's surface, Microwaves can either be absorbed or reflected out into space by the Ionosphere and Radio frequencies reach earth's surface and enable earth-to-space communication.

UV, VIS and IR frequencies are thought to be key components in human evolution and have had substantial impact on human physiology, as they impact our circadian rhythm, and overall wellness. The amount of UV, VIS and IR light we experience depends on the time of date, our latitude, altitude and time of year. The light frequencies change throughout the day within a particular latitude, altitude and time of year, and the main reason for this is the sun's elevation. At sunrise, the VIS and IR light are present while UV is not and stay present until sunset. When the sun elevation is greater than or equal to twenty degrees UVA light appears, when the elevation is greater than or equal to thirty degrees all light frequencies are present, when sun elevation drops below thirty degrees UVB light disappears and when sun elevation drops below twenty degrees UBA light disappears. All of these different light frequencies interact with our eyes and body.

Circadian rhythm is one of the body changes that is greatly influenced by the available light frequencies. Our natural Circadian rhythm dictates that our light and dark cycles are synchronised with the sun and the earth, and health issues may arise when this doesn't happen. Circadian rhythm is the organisation of our molecular and quantum processes in the human body in response to the environment including the natural world. This also includes the control of our gene expression, biochemical reactions such as hormonal changes, tissue and organ growth and repair, Hypothalamus, the Suprachiasmatic nucleus and the eye.

One of the main hormones that is greatly influenced by light is Melatonin. Melatonin is very important and is used for more than just sleeping. Melatonin beings with the interaction of UV light and the amino acid Tryptophan in the eye, which then catalyses the conversion of Tryptophan into Serotonin. Serotonin is then stored in the gut and is released at night from the gut and made available to the Pineal Gland. From there it is converted to Melatonin in the Pineal Gland. For this natural process to happen the Melanopsin and Neuropsin receptors in the eye and the Retinal Pigment Epithelium must not detect blue light after sunset, and digestion should be absent in the gut at night. Apart from sleep, Melatonin also reduces body temperature and reverses the DC electric current in the brain.

Different light frequencies eventually cause the release of Growth Hormone, Thyroid Stimulating Hormone, Adrenocorticotropic Hormone, Follicle Stimulating Hormone, Melanocyte Stimulating Hormone, Luteinizing Hormone and Prolactin. Light also impacts the Pituitary Gland which functions as a bridge that also includes the Hypothalamus and works to connect the Autonomic Nervous System to the Endocrine and the Immune systems. To help wake us up and counter Melatonin, Cortisol is naturally triggered as a healthy response. However, the function of Cortisol can change when in stressful situations. Natural morning sunlight, assists the body in creating Vitamin D. In order for this to happen we need approximately forty-five percent skin exposure when the sun is at approximately forty-five degrees. If this is done successfully, the body will produce Vitamin D which serves many functions including – helping the kidneys to hold onto as much of the Melatonin Sulphate as possible, promoting a health gut and helping the body to improve and maintain healthy Bones, Intestines, Immune system, Cardiovascular system, Pancreas, Muscles, Crain and control of Cell Cycles. A lack of sunlight may also cause a phenomenon known as Sundowning which has several negative impacts on the body including decreases in melatonin in the body, increases in agitation, confusion and anxiety in the late afternoon and evening, deterioration of the Suprachiasmatic Nucleus (SCN) and may worsen symptoms of or increase risk of dementia.

Due to many causes and scenarios in the modern human world, artificial light is often used and overused. Whilst this serves many positive functions, it can have a negative influence on the human body including – altering electron and proton flows to the Mitochondria, negatively impacting gene expression, leading to Leptin or Insulin resistance, lowering Oxytocin resistance, increase the cravings for carbohydrates, increase dehydration, and increasing the risk of Cataracts, Diabetes, Anxiety, Depression and Multiple Sclerosis. Cortisol is also often impacted by the modern world, but also plays a secondary role in assisting the circadian rhythm in waking us up.

Regarding the earth's impact on our individual health, there is a phenomenon known as the 'Schumann Resonance'. The Schumann Resonance is the negative electrical charge of the earth, with a frequency of 7.83. The surface of the earth is electrically conductive, possessing a continuously renewed and limitless supply of free electrons. The earth's surface and the ionosphere form a resonating cavity charged by broadband electromagnetic impulses such as lightning strikes. This results in a negatively charged surface of the earth known as the Schumann Resonance. The easiest way to access this helpful charge is to make contact with the earth directly for at least fifteen minutes. This may lead to several benefits including – the reduction of electric fields induced on the body, reduction of overall stress levels and tension, improvement in the Autonomic Nervous System, and may be anti-inflammatory. In the modern world however, there are many Non-Native Electromagnetic Fields (EMF) which can impact our health negatively such as increasing our risk of disease. The low frequency examples include – house wiring, lighting and microwaves; and the radio frequencies include cordless phones, mobile phones, Wi-Fi, etc.

Resonance is also discussed by professionals who identify themselves as holistic healers when discussing healing through sound. Healing Sounds have been used by many cultures for thousands of years, from Ancient Egyptians, Native American tribes, Indian cultures, Ancient Greek cultures, and more. The "western" scientific approach started investigating healing sounds in the early to mid-1800's. It is thought that humans have a natural resonance (of 9-16 Hz), and that our emotions also influence this natural resonance. Sound healing is theorised to work by influencing these resonances to help a person relax, focus, develop insight, improve sleep, lessen pain and for other. There are different types of sound healing including Sound Baths, Guided Sound Meditations, Chanting, Vibroacoustic, Acutonics and Binaural Beats. As a result, there are many different tools that people and sound healing practitioners use including Tibetan Singing Bowls, Crystal Singing Bowls, Tamburra, Bamboo and Native American Flutes,

Bells, Steel-Tongue Drums, Gongs, Rattles, Shakers, Frame Drums, Chimes, Tuning Forks and Rainsticks.

I will only briefly discuss the immune system, from the allergy perspective, here as it is covered in various other chapters in this book. An allergy is an over-response of the body's immune system, in its effort to protect the body against what it perceives to be a threat. When the immune system's resistance is down, sensitivities are increased, and allergies develop. There is a chemical in the body known as Histamine. Histamine is released by white blood cells called Basophils, the blood platelets and the mast cells in the area of the body a foreign substance is residing. The Histamine's jobs include dilating the blood vessels so that extra lymphocytes can arrive quickly, and speeding up the metabolic rate of all the cells in the area so they can have the energy to protect themselves from foreign bodies and defeat the foreign bodies. Most of the time this is helpful, and causes only minor inflammation, however when there is an allergic reaction too much histamine is released and the body over-responds to certain antigens causing an allergen (and as a result an allergic reaction). Causes can include food, poor mental health (prolonged stress etc. can cause an autoimmune response), stings, environment (dust, grass, etc.), overexposure at young age, genetic expression, migration of allergies to different body locations and body systems, medical conditions, addictions (to substances, pain, foods, allergen producing substances), and an overdosing of vitamins and medication and other causes. Lifestyle choices such as food and exercise can often assist with minor allergies, however, doctors, dieticians, psychologists, naturopaths, and other trained professionals can all be helpful. Whilst medical approaches should normally be step one of the person's journey, other approaches such as finding and treating the cause of the allergy should also be attempted with a trained professional.

Pilates Understanding

Pilates was Invented by Joseph Pilates after he overcame his own health and medical difficulties by self-study including learning from 'Eastern' and 'Western' anatomy and fitness theories and paradigms both ancient and modern. During World War one he refined his skill and understanding of the human body and after various jobs in fitness, martial arts and other areas, he moved to the United States of America. In 1925 he met his partner Clara and together they created 'Contrology' and following his death in 1967, Joseph and Clara's teaching method became known as 'Pilates'. Clara's work with the human body and fitness following Joseph's death, allowed for non-athletes to benefit from the 'Pilates Method' due to her own nursing experience and her different teaching ability and method. It is often suggested that Clara's ability to teach the complex movements and theory, to untrained athletes is what allowed for the popularity of the Pilates Method to spread.

Today's Pilates Method continues the previous work, and focuses on 'Intersegmental Stabilisation', otherwise known as 'Core Stability'. The simplest way to explain Pilates to non-practitioners is through the term "Strength through Core". The three fundamental principles of the original Pilates philosophy are 'Whole Body Health', 'Whole Body Commitment' and 'Breath'. The principles of Modern Pilates are Relaxation, Concentration, Alignment, Centring, Breathing, Coordination, Flowing Movements and Stamina, with the terms "muscle control" and "body awareness" encompassing many of these elements. In Pilates, Core Stability, is more than just a six pack, it includes the joint and muscle health along the spine, otherwise known as spinal stability and can include retraining movement patterns.

Relaxing helps the person begin their body-mind condition journey which then leads to a better awareness of the body. This helps a person alleviate some of their physical expression of stress whilst lengthening and strengthening their body. Concentration follows on from the concepts of Relaxation and furthers this through use of visualising which helps the build the neural networks in the brain-body connection journey.

Alignment is the concept of the whole body being involved with every movement, where one movement or alteration will affect the whole body. This also assists with Postural Balance and enhance the overall sense of wellbeing. Centring refers to the "Powerhouse" meaning core stability, and this is achieved through holding the pelvis in neutral position whilst activating the Transverse Abdominals. Pilates Breathing emphasises the use of breathing to help keep the bloodstream "pure" during exercise, which oxygenates the blood and eliminates noxious gases. It is believed that breathing in the Pilates way, helps to expand the thoracic muscles and the ribs to enable the lungs to help the person reach their full potential.

Pilates has the aim to improve a person's Coordination through guided experimentation of exercises. Flowing Movements in Pilates refers to the "flowing motion outwards from a strong centre". In Pilates it is believed that movement is a sign of life, and that flowing movement gives the sensation of harmony and balance. The final principle in Pilates is Stamina. Stamina in Pilates refers to the gradual refining of the core muscles, through increasing the number of repetitions and/or building on reducing the resistance of the exercises, dependent on the person's goal.

Pilates also looks at Motor Control and the relevant Local Stabilisers and Global Mobilisers. Motor Control is now used when speaking of spinal stability as well as other areas of musculoskeletal stability.

Motor Control is the balance between movement and stability and includes our Central Nervous System (CNS creates a stable foundation for movement of the extremities through co-contraction of particular muscles). Pilates focuses on understanding and improving the Local Stabilisers and the Global Mobilisers in most areas of the body, especially when discussing Motor Control. Local Stabilisers in Pilates refers to the muscles close to the joint, are often postural, are predominantly slow twitch, are type two muscles; whereas Global Mobilisers in Pilates refers to the muscles that are superficial, phasic, are fast twitch and type one muscles.

Contemporary fitness and exercise programs often include Pilates or Pilates inspired exercises as part of their programming due to the overall body benefits. The many progressions and regression in most of the Pilate's exercises make it an accessible form of fitness for the majority of the population. This combined with the idea that it can be taught very differently also allows for different types of fitness goals from cardiovascular fitness to rehabilitation to strength related goals, to be achieved. There will many Pilates and Pilates inspired exercises in the final chapter of this book.

Herbalist Understanding

The majority of the information in this section will be drawn from the Advanced Master Herbalist course I completed but will also contain information from other sources. As with most of these sections, I will only be including information that I found interesting, helpful and or relevant.

There are a lot of different herbs that can be used to help a person recover from an injury or illness or maintain better overall health. Herbs can have multiple effects and can be used as the primary or as support ingredients, depending on the goal or problem. Some types of effects herbs can have include Adaptogen, Adjuvant, Alterative, Amphoteric, Anaesthetic, Analgesic, Anti-Allergy, Anticatarrhal, Anti-Emetic, Anti-Inflammatory, Antimicrobial, Anti-Rheumatic, Antispasmodic, Antitussive, Astringent, Bitter, Calming/Relaxing, Cardioactive, Cardioprotector, Cardiotonic, Cholagogue, Decongestant, Demulcent, Diaphoretic, Diuretic, Emmenagogue, Expectorant, Gut Healing, Hepatic, Hypnotic, Hypotensive, Immunomodulator, Immunostimulant, Lymphatic, Nervine, Synergistic (pairing), Warming, and others. Some herbs can only be used short term, some are safe to use long term, some herbs are known as restricted herbs due to either their potency or possible side effects and herbs differ in how quickly they assist the body.

As herbs have so many possible healing abilities, it is no surprise that there are many physical and mental health ailments and difficulties that can be supported by or treated with herbal remedies. I have focused primarily on the herbal effects as there is too much information regarding herbal use to fit in this type of book. Herbs have been used for thousands of years, and research from at least the last fifty years has found that they may be useful for various mental health presentations such as PTSD, Anxiety, Depression, Sleep, Cognitive Health, Physical Recovery. I have tried not to focus on too many conditions directly as I believe this information is useful to know, but not a replacement for trained medical advice and knowledge. It is also important to remember that some herbs may be contraindicated with prescribed medication, or may other side effects, so check with a trained herbalist/dietician (who has extra training in herbs) prior to consuming medication and herbs simultaneously. Some herbs are best used when fresh and many are useful when dried. There are many ways in which to use herbs including sprays, tinctures, lozenges, syrups, creams, teas, infusions, capsules, baths, salves, suppositories, recipes and more. However, I have not discussed this in the book as I have never practiced as a herbalist and felt I would not be doing the herbal practice justice if I attempted to explain the processes in which I only have a theoretical understanding of. Below are some of the various effects of herbs including example for each of the effects.

Adaptogen herbs can be used to help the body manage or adapt to a chronic condition, are known as metabolic regulators and are often used as a secondary effect with many other herbal remedies. Other roles Adaptogen herbs can assist with include – helping to address debility and potential true exhaustion (adrenal burnout), supporting the body to manage prolonged pain, assisting with managing severe stress, assisting in the treatment of Fibromyalgia, supporting the immune system and helping to guard against allergic reactions, assisting the digestive system, assisting in management of gastrointestinal pain, assisting in management of insomnia and anxiety, assisting the anti-inflammatory, antioxidant, antiviral, expectorant and demulcent properties of other herbs and foods, reducing the allergic response to stress, assisting with liver health, assisting energy levels, supporting thyroid recovery and used as a primary or secondary herb in general and adrenal tonics. There are many examples of adaptogen herbs including Ashwagandha, Astragalus, Brahmi (Bacopa Monnieri), Codonopsis (Dang Shen), Damiana, Ginseng, Gotu Kola, Guduchi, Hawthorn Blossom, Holy Basil, Hops, Korean (Panax) Ginseng, Liquorice, Passionflower, Nettle Seed, Rehmiana, Reishi, Rhodiola, Saw Palmetto, Schisandra, Shatavari, Shiitake, Siberian Ginseng, Tulsi, and Valerian.

Adjuvant herbs are used to enhance the production of antibodies to enhance the immune response, are used to increase the efficacy of the blends they are in (acting like an accelerator), promoting movements of the constituents from the digestive tract into the bloodstream (to enhance absorption), and some act via the stimulation of the nervous system and circulation. The many examples of Adjuvant herbs include – Andrographis, Angelica, Ashwagandha, Astragalus, Bayberry, Cayenne, Cinnamon, Damiana, Echinacea, Ginger, Panax Ginseng, Picrorrhiza, Poke Root, Prickly Ash, Rosemary, Shatavari, Wood Betony and Yarrow.

Alterative herbs are used – to enhance the metabolic and eliminatory processes, when a person's body is out of balance (often due to chronic disease, stress, inflammation or autoimmune diseases), for skin conditions, to help address toxicity, to slow the degeneration of joints, to help cleanse the body of toxins and to support the eliminatory organs. Alterative herb examples include – Birch, Black Cohosh, Bogbean, Burdock, Celery Seed, Cleavers, Dandelion Leaf and

Root, Echinacea, Figwort, Fringetree, Fumitory, Garlic, Golden Seal, Mahonia (Oregon Grape), Nettle, Poke Root, Red Clover, Sarsaparilla, Uva-Ursi and Yellow Dock.

Anaesthetic herbs are often used to help numb the throat. They can be used in various forms including a spray, syrup, gargle and lozenge form. Example herbs include Clove, Peppermint, Propolis and Myrrh.

Analgesic herbs are mostly used to manage pain, and this includes – General pain relief, easing of diarrhoea related pain, headaches and aching muscles, lessening of muscular tension around the joints, managing stress and as external warm compresses around the ear and jaw. Examples of Analgesic herbs include Belladonna, Californian Poppy, Cramp Bark, Datura, German Chamomile, Henbane, Hops, Jamaican Dogwood, Lavender, Meadowsweet, Passionflower, Peppermint, Valerian, Wild Lettuce and Yellow Jasmine.

Anti-Allergy herbs are used to support the immune system's response to allergens either as primary or secondary herbs and can be used internally or externally. They are used as they help to dampen the immune response and it is theorised, their action on the Mast Cells supports the immune system. They are also useful when vulnerable populations such as young children or elderly adults have a respiratory tract infection or digestive infection. Example herbs with this action include – Albizzia, Baikal Skullcap, Feverfew, German Chamomile, Lemon Balm, Nettle, Plantain, Wild Yam and Yarrow.

Anticatarrhal herbs are often used when there is an excess of mucous, associated with the inflammation of the surround mucous membranes. The mucous is often located in the back of the nose, throat or sinuses and occasionally can be found to impact bowel movements. These herbs often are used as they either help to dry the mucous or make the mucous watery (through a warming process) to allow for it to be expelled more easily. Examples of these herbs include Cayenne, Cranesbill, Echinacea, Elderflower, Elecampane, Eyebright, Garlic, Golden Rod, Goldenseal, Ground Ivy, Herb Robert, Hyssop, Irish Moss, Marshmallow (Flowers, Leaves and Root), Mullein, Peppermint, Plantain Ribwort, Sage, Thyme, Uva-Ursi, Volatile Oils and Yarrow.

Anti-Emetic herbs are helpful to manage nausea and when the digestive system is infected by mucous dripping into the system. Example of helpful herbs include – Aniseed, Black Horehound, Fennel, Ginger, Henbane, Lobelia, Meadowsweet, Peppermint and Spearmint.

Anti-Inflammatory herbs have three main classes – Volatile Oils, Salicylates and Steroid Precursors. Herbs that contain volatile oils often help to regulate inflammation via inhibition of the formation of certain Leukotrienes and via the inhibition of the peroxidation of Arachidonic Acid. These are often better suited for managing inflammation in the respiratory system and digestive tract and occasionally are used for external inflammation.

Salicylate herbs are best used for inflammation in the joints and for inflammation across the musculoskeletal system when injuries or infections occur. Inflammation can also occur follow long periods of physical tension in the body or long periods of spasmodic actions that may lead to injuries. Steroid Precursor herbs often are used to help the body to make steroidal compounds to manage inflammation such as in autoimmune disease inflammation. Some herbs can have multiple anti-inflammatory pathways they assist with.

Some examples of Anti-Inflammatory herbs include –
Angelica, Burch, Black Cohosh, Black Haw, Blue Cohosh,
Bogbean, Celery Seed, Chickweed, Cleavers, Cornsilk, Couch
Grass, Cramp Bark, Cranesbill, Devil's Claw, Dill, Elderflower,
Fenugreek, German Chamomile, Golden Rod, Goldenseal,
Hawthorn, Herb Robert, Horse Chestnut, Irish Moss, Lady's
Mantle, Lavender, Lemon Balm, Lime Blossom, Liquorice,
Marigold, Marshmallow (root, leaves and flowers),
Meadowsweet, Mullein, Nettles, Peppermint, Plantain
(Ribwort), Sage, Shepperd's Purse, Slippery Elm, Spearmint,
Wild Yam and Yarrow.

Antimicrobial herbs are often used as secondary herbs
depending on the type of infection and if they are being used
internally or externally, and they can be used in most if not all
body systems. Example of when they are used as primary herbs
include damage in the digestive mucosa that leads to infection
and to support specific types of infection. Often, they are used
as secondary herbs to assist with other herbal actions including
– Anti-inflammation, Anti-septic, Antispasmodic, Astringent,
Bitter, Carminative and Immune related actions. Examples of
these herbs include Aniseed, Bearberry, Caraway, Cayenne,
Cinnamon, Clove, Cranberry, Echinacea, Elecampane,
Elderberry, Garlic, Gentian, German Chamomile, Ginger,
Goldenseal, Hops, Juniper, Lemon Balm, Liquorice, Mahonia
(Oregon Grape), Marigold, Marjoram, Myrrh, Pau d'arco,
Peppermint, Pine, Plantain (Ribwort and Greater), Rosemary,
Sage, St John's Wort, Thyme, Usnea, Uva-Ursi, Wild Indigo,
Wormwood and Yarrow.

Anti-Rheumatic herbs are used to help treat the
conditions of heat, pain and swelling in the joints. These herbs

have various effects including Anti-inflammatory, Promoting blood flow, Diuretic, Antispasmodic, Alterative and other effects. Useful herbs from this section include Angelica, Bayberry, Bearberry, Birch, Black Cohosh, Bladderwrack, Blue Cohosh, Bogbean, Burdock, Cayenne, Celery Seed, Cleavers, Cramp Bark, Dandelion (Leaf and Root), Devil's Claw, Feverfew, Ginger, Horseradish, Juniper, Mahonia (Oregon Grape), Meadowsweet, Mugwort, Mustard, Nettle, Parsley, Poke Root, Prickly Ash, Rosemary, Wild Yam, Willow, Yarrow, and Yellow Dock.

Antispasmodic herbs are often used to relax the musculature by lessening the spasms in muscle fibres via the autonomic nervous system. This doesn't affect the Central Nervous System but does help to ease pain. Examples of herbs with this effect include Angelica, Aniseed, Belladonna, Black Cohosh, Black Haw, California Poppy, Caraway, Catnip, Celery Seed, Chery Bark, Cramp Bark, Damiana, Datura, Dill, Elderflowers and berries, Fennel, Fenugreek, Feverfew, German Chamomile, Greater Celandine, Ginger, Henbane, Hops, Hyssop, Lavender, Lemon Balm, Lime Balm, Liquorice, Lobelia, Marjoram, Motherwort, Mugwort, Passionflower, Red Clover, Rosemary, Sage, Skullcap, St John's Wort, Thyme, Wild Carrot, Wild Lettuce, Wild Yam, Valerian and Vervain.

Antitussive herbs help to reduce coughing which can lead to improvements in sleep and lessening of pain with Wild Cherry Bark being the most recommended herb.

Astringent herbs are used for a lot of different reasons. They are used externally to help stem bleeding, to help recovery of broken and weeping skin, to reduce inflammation, to help tone veins and to make the skin more receptive to other herbs (most useful when there are open scars and spots that have been recently squeezed). When Astringent herbs are used internally, they can assist by reducing secretions from mucous membranes, heal the bladder wall, act as an anti-diarrhoeal, bind tissue to improve tissue integrity and reduce inflammation. Astringent herbs are recommended for shorter term use only. Examples of Astringent herbs include Agrimony, Bayberry, Cranesbill, Elderflower, Eyebright, Hazel, herb Bennett, Herb Robert, Lady's Mantle, Meadowsweet, Nettle, Oak, Plantain, Shepperd's Purse, Tormentil, Witch Hazel and Yarrow

Bitter herbs have an active ingredient that stimulates our taste buds leading to reactions in our digestive system, preparing the body to digest food. Bitter herbs also have various other uses in the body including increasing appetite in people who don't have one, supporting the liver, maintaining body sugar levels, helping to manage Leaky Gut Syndrome, assisting to fight inflammation in the body, helping the body to fight infections in the digestive system, assisting the body's skin health, helping the body during convalescing stages of colds and flu, strengthening the person's digestion to ensure the body gets the best nutrition and supporting the liver during hormonal fluctuations. Herb examples include Agrimony, Angelica, Bearberry, Bogbean, Catnip, Centaury, Dandelion Root, Gentian, German Chamomile, Ginger, Golden Seal, Lemon Balm, Mugwort, Purslane, Sage, Spearmint, Thyme, White Horehound, Wormwood and Yarrow.

Cardioactive herbs are often for short term use only as they contain Cardiac Glycosides which can have a negative impact on mental and physical health if used for too long. Cardioactive herbs have many uses including promoting the urine to assist with Oedema, inhibiting the reabsorption of water in the kidney and promoting blood flow to the kidney. Herb examples include Broom, Bilberry, Cayenne, Figwort, Garlic, Ginger, Ginkgo, Hawthorn Berries and Flowering Tops, Horse Chestnut, Lemon Balm, Lily of the Valley, Lime Blossom, Motherwort, Prickly Ash, Rosemary and Yarrow.

Cardioprotector herbs can be used longer term, although short term use is still recommended. These herbs are used to assist in the management of hypertension, to promote the production of urine to assist with Oedema, inhibit the reabsorption of water in the kidney and to promote blood flow to the kidney. Herb examples include Broom, Bilberry, Cayenne, Figwort, Forskohlii, Garlic, Ginger, Ginkgo, Hawthorn Berries and Flowering Tops, Horse Chestnut, Lemon Balm, Lily of the Valley, Lime Blossom, Motherwort, Prickly Ash, Rosemary and Yarrow.

Cardiotonic herbs can be used longer term, although short term use is still recommended. These herbs are used to strengthen the cardiovascular system, promote healthy elasticity of the system, promote the production of urine to assist with Oedema, inhibit the reabsorption of water in the kidney and to promote blood flow to the kidney. Herb examples include Broom, Bilberry, Cat's Claw, Cayenne, Figwort, Forskohlii, Garlic, Ginger, Ginkgo, Guggul, Hawthorn Berries and Flowering Tops, Horse Chestnut, Lemon Balm, Lily of the Valley, Lime Blossom, Motherwort, Prickly Ash, Rosemary and Yarrow.

Cholagogue herbs are used to promote the flow of bile from the liver and to assist the digestive system. Cholagogue herb examples include Artichoke, Barberry, Dandelion Root,

Fringe Tree, Fumitory, Gentian, Goldenseal, Greater Celandine, Lemon Balm, Mahonia (Oregon Grape), Rosemary, Sage, Wild Yam and Yellow Dock.

Decongestant herbs are often paired with Anticatarrhal herbs due to their overlapping effects. Decongestant herbs are used to – assist the respiratory system, relieve excessive mucous and catarrh in the upper respiratory tract, manage sinusitis, manage blocked ears and hay fever, assist the body in the loosening of the mucous through liquefying the mucous, reduce the secretions from the mucous membranes and to maintain healthy skin. Herb examples include Aniseed, Elderflower, Garlic, Ground Ivy, Eyebright, Plantain, Golden Rod, Goldenseal and Mullein.

Diaphoretic hers are used to assist the immune system and help to manage fever. Herb examples include Catnip, Cinnamon, Elderflower, Ginger, Horseradish, Lime Blossom, Peppermint and Yarrow.

Diuretic herbs have a side effect of removing electrolytes from the body which needs to be observed due to possible impacts on muscles including the heart. Diuretic herbs assist the body in various ways including – promoting the production of urine to assist with Oedema, inhibiting the reabsorption of water in the kidney, promoting blood flow to the kidney, flushing the urinary system to remove infections, encourage other herbs to make contact with the mucous membranes, remove excess fluid from the body (helping the cardiovascular system, menstrual cycle and lessening congestion with the reproductive organs) and removing waste products to lessen the chance of waste excreting through the skin. Herb examples include Agrimony, Bearberry, Birch,

Burdock, Broom, Buchu, Celery Seed, Cleavers, Cornsilk, Couch Grass, Dandelion Leaves, Elderflowers and Berries, Hawthorn, Horsetail, Juniper, Lily of the Valley, Lime Blossom, Parsley, Pellitory of the Wall, Saw Palmetto, Uva-Ursi, Wild Carrot and Yarrow.

Emmenagogue herbs have been found to be very dangerous for pregnancy, so as always, medical advice is recommended. Emmenagogue herbs assist with stimulating menstruation, reducing excessive bleeding during menstruation, assist in regulating hormones and can be used to strengthen the uterus in preparation for trying to conceive and prepare for labour. Herb examples include Agnus Castus, Black Cohosh, Black Haw, Blue Cohosh, Condurango, Cramp Bark, Fenugreek, Feverfew, Gentian, German Chamomile, Ginger, Goldenseal, Hyssop, Lavender, Lime Blossom, Marigold, Motherwort, Mugwort, Nasturtium, Parsley, Partridge Berry, Pasque Flower, Pennyroyal, Peppermint, Poke Root, Raspberry Leaf, Rosemary, Rue, Sage, Southernwood, Tansy, Thyme, Valerian, Vervain, White Horehound, Wormwood and Yarrow.

Expectorant herbs have several types of effects including Relaxing, Stimulating and Amphoteric, with secondary types of Expectorant herbs including soothing Demulcent and Aromatic Expectorant herbs. Relaxing Expectorant Herbs can include Demulcent, Antispasmodic and Anti-inflammatory herbs that – are high in volatile oils, high in mucilage that can exert a demulcent action on the gut (and respiratory system), help reduce spasm in the bronchial tubes and make mucous watery (so it's easier to expel) to help a dry congested cough. Examples of Relaxing Expectorant herbs include – Aniseed, Cherry Bark, Goldenseal, Grindelia, Hyssop, Irish Moss, Lobelia, Liquorice, Lungwort, Marshmallow

(Flowers, Leaves and Root), Thyme and Vervain.

Stimulating Expectorant herbs – helps to thin mucous and irritate bronchioles to encourage coughing, are used to clear lungs of large amounts of mucous and congestion (such as bronchitis and chest coughs) and are high in saponins, alkaloids and volatile oils. Examples of herbs include Anise, Caraway, Cowslip, Daisy, Elecampane, Fenugreek, Ipecac, Primrose Root, Sweet Violet and White Horehound.

Amphoteric Expectorant herbs can be stimulating and relaxing and are often used in addition to other treatments. Examples include Mullein, Elderflower, Elder Berries and Garlic.

Other ailments that the various types of Expectorant herbs can assist with include – lingering colds, immune support and in various paediatric conditions. Other Expectorant herbs include – Fennel, Lobelia and Sundew.

Gut Healing herbs are used to – assist the body in managing inflamed bowel conditions, helping the digestive system's interaction with the immune response, assist in managing leaky gut issues and autoimmune issues, assisting in combating malnutrition and dysbiosis, address intestinal damage and assisting the body in managing allergies that may stem from the gut. Herb examples include Aloe Vera, Agrimony, Bayberry, D'Arco, German Chamomile, Liquorice, Marigold, Marshmallow Root, Meadowsweet, Plantain, Slippery Elm, Wild Yam and Yarrow

Hepatic herbs have many uses including promoting healing and regeneration in the liver, supporting the liver in its detoxification and eliminatory processes by stimulating bile production and bile flow, managing long term constipation, assisting the digestive system's interaction with the body's immune response, supporting the liver during hormonal fluctuations, supporting the liver in the breakdown of hormones and supporting the liver during the menstrual cycle. Herb examples include Agrimony, Artichoke, Barberry, Burdock Root, Celery Seed, Centaury, Cleavers, Dandelion Root, Elecampane, Fennel, Fringetree, Fumitory, Gentian, German Chamomile, Goldenseal, Lemon Balm, Mahonia (Oregon Grape), Milk Thistle, Motherwort, Prickly Ash, Schizandra, St John's Wort, Turmeric, Wild Yam, Yarrow and Yellow Dock.

Hypnotic herbs contain alkaloids and are normally used to promote sleep. Other reasons they may be used include managing stress, anxiety, panic or over-stimulation by encouraging the parasympathetic system, encouraging deep sleep, assisting in the management of insomnia and supporting the nervous system by assisting the person to manage the side effects of autoimmune related diseases such as Rheumatoid Arthritis. Herb examples include California Poppy, German Chamomile, Hops, Lime Blossom, Motherwort, Mugwort, Oats, Passionflower, Skullcap, St John's Wort, Valerian, Vervain, Wild Lettuce and Wood Betony.

Hypotensive herbs are used to help manage high blood pressure by relaxing the muscles that circle the blood vessel, calcium channel blocker activity and via the ACE inhibitor activity of certain procyanidins and flavonoids. Herb examples include Black Cohosh, Black Haw, Blue Cohosh, Cramp Bark, Fenugreek, Forskohlii, Garlic, Hawthorn, Lime Blossom, Mistletoe, Motherwort, Nettle, Passionflower, Siberian Ginseng, Skull Cap, Valerian, Vervain, Yarrow and Yellow Jasmine.

Immunomodulator herbs assist the immune system by – stimulating and addressing imbalances in the immune system, modulating the immune system's response, normalising the immune system response longer term, stimulating and suppressing aspects of the immune, enhancing the activity of T-suppressor cells (which may assist in combating tumours) and possibly stimulating the body's killer cells and macrophages. Herb examples include Aloe Vera, Andrographis, Ashwagandha, Astragalus, Baikal Skullcap, Brahmi, Cinnamon, Codonopsis, Dong Quai (Chinese Angelica), Echinacea, Garlic, Gotu Kola, Holy Basil, Marshmallow, Pau d'arco, Plantain, Poke Root, Rehmannia, Reishi, Rhodiola, Shiitake and Thyme.

Immunostimulant herbs help to stimulate the immune system if a person is debilitated due to chronic health issues or other severe mental or physical issues such as exhaustion. These herbs also help the body to fight infection (especially in the upper respiratory tract), assist the body in non-specific stimulation of the immune system, assist in combating flus, and help people break the cycle of poor recovery. Herb examples include Aloe Vera, Andrographis, Astragalus, Brahmi, Cinnamon, Codonopsis, Dong Quai (Chinese Angelica), Echinacea, Garlic, Gotu Kola, Holy Basil, Marshmallow, Pau d'arco, Plantain, Poke Root, Rehmannia, Reishi, Rhodiola, Shiitake, Thyme and Wild Indigo.

Lymphatic herbs' main role is to assist the body in draining/channelling away fluid and preventing too much storage of flood. Lymphatic herbs are also used to – clean the fluid and interact with the immune system by neutralising potentially harmful particles and microbes, helping the immune system through assisting the lymph nodes when they are swollen and tender, helping the tonsils when they are inflamed, red, swollen or congested, helping to clear congestion around the mucous membranes and supporting the lymphatic system in managing inflammation. Helpful herbs include Bladderwrack, Astragalus, Cleavers, Burdock, Echinacea, Figwort, Liquorice, Panax Ginseng, Poke Root, Marigold and Wild Indigo.

Nervine herbs are used to manage physical illnesses caused by high levels of stress, tension and other mental health difficulties such as anxiety. Some Nervine herbs also come under their own section known as Anxiolytic (used to treat anxiety). They are used to relax the nervous system, strengthen the nervous system and address the stress causes of illnesses such as IBS. Other uses include improving sleep, lessening panic attacks, assisting in wound healing, assisting the immune system, managing restlessness, assisting in management of convulsions in epilepsy, lessening tension headaches, lessening stress related hiccups and assisting in management of depressive symptoms. Research so far, has found that they don't suppress the REM phase of sleep (whereas many pharmaceutical sleep aids do), are rarely addictive and rarely interrupt natural body brain communication. Herb examples include Brahmi (Bacopa Monnieri), Black Cohosh, Blue Cohosh, Black Haw, Black Horehound, California Puppy, Catnip, Cramp Bark, Damiana, German Chamomile, Ginkgo Biloba, Gota Kola, Holy Basil, Hops, Kava, Lavender, Lemon Balm, Lemon Grass, Lime Blossom, Lobelia, Meadowsweet, Motherwort, Mugwort, Oats, Passionflower, Pennywort, Red Clover, Rose, Rosemary, Siberian Ginseng, Skull Cap, St John's Wort, Valerian, Wild Lettuce, Wild Oats and Wood Betony.

Synergistic (pairing) of herbs refers to herb pairing that enhance each other's effects, and this is referred to as working synergistically. Synergistic pairs include – Cleavers and the Viola Species, Agrimony and most other herbs that act to assist the digestive system, Angelica and White Horehound & Elecampane in the respiratory system, Black Pepper & Turmeric, by significantly Enhancing its Absorption in the gut, Marshmallow and Solomon's Seal & Slippery Elm for activity on the Joints, Prickly Ash and Yarrow by Enhancing Yarrow's effects on the Capillaries and Peripheral Circulation, and Skullcap and Passionflower by enhancing each other's Anxiolytic and Nervine Effects.

Everyone including professionals have their favourite herbs for various presentations. However, some research suggests there are herbs that have high accessibility, multi-use and have the most research regarding effectiveness and safety. According to some research the top five herbs for Mental health are Bacopa, Rhodiola, Kava, Liquorice and St. John's Wort; and the top five herbs for overall physical health are Turmeric, Chamomile (preferably German Chamomile), Milk Thistle, Basil and Cinnamon. Alongside all the above herbalist uses and options there is a supplementary practice known as Aromatherapy. Aromatherapy is often used alongside herbalist practices and often involves using essential oils (volatile oils) to engage aroma (smell) to help alter consciousness to alter mental and physical health. One theory on how aromatherapy may work suggests that the binding of the chemical compounds in the essential oils may bind or interact with the olfactory bulb, which then in turn, triggers the brain's limbic system. As a result, there are many benefits, however, there may be a mental health related risk if the person has previous traumas associated with smell, the aromatherapy may accidently trigger this, even if the person is unaware. Some benefits of using aroma therapy either as its own practice or alongside other approaches include – bringing pleasure or influencing mood, reducing pain, assisting sleep and relaxation, assisting focus and learning, influencing perception of self and of own health, it may assist as an antibacterial (through cleaning) and as an anti-inflammatory. Aromatherapy can be mixed with other liquids such as carrier oils (coconut or almond oil), diffused into a mist, made into a plaster on the skin or directly inhaled when mixed with heated water.

Nutrition

This section will focus on how nutrition directly and indirectly impacts our health. Nutrition is a very large and open-ended area of research and understanding. As such, the majority of information presented here, will be focusing on its impacts on mental health whether directly or indirectly.

There are many different types of diets/food plans that people around the world follow, and often what is generalised advice won't work for everyone. Each person has their own dietary needs, and some practiced diets include Ancestral, Atkin's Carnivore, Paleo, Modified Carnivore, Ketogenic, Anti-Inflammatory, Gut and Psychology Syndrome (GAPS), Mediterranean, Raw, Vegetarian, Ayurvedic, Chinese Five Element, Macrobiotic, Vegan, Fasting and Detoxification. The individual's genetic difference means that our biochemistry is the main determinant about what foods/fuel is better for a person, and the rate of glucose oxidation is the rate at which our body burns fuels. This means that some nutrition plans may be harmful or counterproductive. Our cultural-genetic heritage determines whether the individual will do better with higher percentages of fats and proteins, or better with less fats and higher percentages of carbohydrates.

The main role of carbohydrates is to provide the body and all its cells with energy. Carbohydrates can be divided into at least five different categories including monosaccharides, disaccharides, oligosaccharides, polysaccharides, and sugar alcohols. The brain and red blood cells rely heavily on carbohydrates being converted into glucose for energy. Fibre is important and can come as either Dietary Fibre or Functional Fibre, although these terms are often used as the same in the general population under the label Fibre. Dietary fibre is the carbohydrates and lignin that are found in plants and are not digested and absorbed in the small intestine. Functional fibre consists of isolated or purified carbohydrates that are not digested and absorbed in the small intestine and have beneficial physiological effects in humans. The main purposes of fibre include laxation, attenuation of blood glucose levels and normalisation of serum cholesterol levels. Proteins form the major structural components of all the cells of the body and function as enzymes in membranes, as transport carriers and as hormones. Amino acids are elements of protein and act as precursors for nucleic acids, hormones, vitamins and other important molecules needed in the body. Consuming fats aid in the absorption of fat-soluble vitamins and carotenoids.

People who benefit more from fats and proteins are known as 'Fast Oxidisers'. Fast oxidisers burn glucose too rapidly and require more protein and fats (such as the Inuit peoples). Fast oxidisers often benefit more from carnivorous diets, animal proteins, fats, purine food, organ meats, sardines, anchovies, fruits, vegetables and few grains (approximately 30% protein, 20% carbohydrates, 50% fats). Fast oxidisers will still often meet their body's glucose needs as they will metabolise glucose from the amino acids found in dietary proteins. People who don't burn glucose rapidly enough and require a higher glucose percentage (such as Indian peoples and people living in Tropical countries) are known as 'Slow Oxidisers'. Slow oxidisers often benefit more from vegetarian and pescatarian nutrition plans, plant proteins, nuts, legumes, fish, eggs and low fats (approximately 25% protein, 60% carbohydrates, 15% fats). The majority of the modern-day population fall under the 'Mixed Oxidisers' heading. Mixed oxidisers are people who, as the name suggests, benefit from a mixed protein-fat-carbohydrate diet to balance the rate of oxidisation (approximately 30% protein, 40% carbohydrates, 30% fats). This information is helpful and can be used to help a person direct their efforts such as cooking to their type.

Many health difficulties arise following extended periods of poor sleep and poor nutrition. This statement is simple, but important to reiterate. In the modern world where life can often be faster, more stressful, and require more of a person, many people look for a quick fix to their difficulties. However, most of a person's individual health difficulties can be managed or assisted through the correct nutrition.

Nutrition related information and tips

This subsection will look at some basic information that is helpful in preparing food and make suggestions about what foods may be beneficial. Following this, different foods, vitamins, and minerals will be briefly discussed.

Preparing fresh food where possible may have mental health benefits and can be used as a relaxation or family building tool. It also allows a person to have more control over what they are consuming. Healthy foods to prepare in kitchen for overall mental health and mood improvement include Organic meats (beef, lamb, chicken), pinto beans, eggs, wild caught or wild canned salmon or tuna, olive oil, coconut fat, sweet potatoes, blueberries, raspberries, lemons, oats, green tea, beets, basil, figs, and bitter green vegetables. Dark Leafy greens and Cruciferous vegetables are best when lightly cooked/steamed (still green and slightly crunchy).

When cooking, it is useful to know that oils have different smoke points (when the oil starts to burn and lose nutrients) so if nutrition is important to a person's cooking process, they can adapt the temperature at which they prepare food. Where possible it is advisable to eliminate additives, preservatives, hormone additives, toxic pesticides and fertilisers on food, consume healthy fats, eliminate refined sugars (including carbohydrates) and eliminate refined fats.

Eating antioxidants should form part of the daily or at least weekly practice as they are the body's protection against free radicals that age us and impact many areas of our physical and mental health. Eating the colours of the rainbow (fruit and vegetables) can help to ensure the body is getting the required fuel as different colours in natural foods mean different pigments. Anthocyanins (blues and purples), Betalains (reds and violets), Carotenoids (deep yellow, reds and orange) and Chlorophylls (greens), means different chemical compounds are found in the different coloured foods. Anthocyanins are represented as red and are found in blue berries and cherries; Carotenoids are represented as Beta-Cryptoxanthin, Beta-Carotene and Lycopene, Beta, Cryptoxanthin which are all precursors for vitamin A, are highly bioavailable, and are essential for eyesight, growth and immune response and can be found in orange rind, egg yolk, apple skin, and mandarins. It is suggested that Carotenoids are useful for digestion, especially after large fatty meals. Chlorophylls are represented in leafy green vegetables, green herbs and other foods and can assist with wound healing, assist with iron and lessen inflammation in the body. Eating a spectrum of food proteins (meats, nuts, eggs, dairy), a spectrum of Carbohydrates (root vegetables, soluble fibres such as psyllium seeds, fruits and vegetables, fresh greens, fermented foods, fruits, and Grains are recommended. Healthy fats and other supporting foods are also beneficial.

When cooking with oil there are several ideas and options to think about. Is the oil stored in a dark green bottle to lessen light exposure damage, if not can you afford to buy an alternative that is. Some research has suggested when oil receives too much light exposure damage, it can lose nutrient quality. Ghee is an optimal cooking oil and butter replacement, and it may assist with absorption of positive effects from herbs and spices. Coconut fats can be an alternative for cooking as it can help with internal and external health difficulties. Sesame oil (raw) in small amounts can be useful as it helps with lessening gum health difficulties which can also assist the body in lessening heart disease and dementia risk.

As well adding or being a base for flavour, "cooking"

herbs can assist in overall health with basil, oregano, cardamon, cinnamon, curcumin, turmeric, ginger and garlic being examples. Sea Salt is often used to naturally enhance flavours, but it is also rich in minerals, supports adrenal function and may help to lessen stress (pink and grey sea salts are recommended. Different vinegars can also assist in flavour and health in cooking with fruit vinegars being the most recommended (such as apple cider vinegar). The health benefits of fruit vinegars can include better management of blood glucose control, increasing levels of vitamin C, antimicrobial, antifungal, boost energy and reduce mild anxiety.

Everyone has been told that eating vegetables are good for you, and there are different types with different benefits. Cruciferous vegetables such as Broccoli, Kale, Brussel Sprouts, Broccoli Sprouts, and Parsley have many health benefits. These include Anti-carcinogenic, Antioxidant, Anti-inflammatory, reversing oxidative stress and improving mitochondrial function, helping to detox body, balancing oestrogen metabolism and helping to release it and assisting the Glutathione production in the body. Nightshade vegetables such as eggplant have many health benefits including fibre, antioxidant and anti-inflammatory. However, they may also cause intestinal permeability and inflammation which may lead to more reflux and pain. For people who have a negative response to nightshade vegetables alternatives such as plantain can be beneficial. Seaweed has been called the vegetable of the sea and has many health benefits. These can include being rich in calcium, phosphorus, magnesium, iron, iodine, sodium, Vitamin C, Vitamin A, Folacin, Vitamin B complex, protein, helping thyroid function, assisting in constipation management and losing weight; and some variations such as brown seaweeds (such as Royal Kombu) help to detoxify heavy metals (such as from air pollution) in the body by binding with the excess metal and radioactive isotypes. Mushrooms can also be beneficial and are sometimes used as meat replacements. Reishi mushrooms are thought to be good for the immune system, Shiitake mushrooms for Vitamin B, Niacin, Choline, Folate, Selenium, Copper, Zinc and Manganese, and Lion's Mane mushrooms for brain and nervous system support, PTSD support, and for improving mild cognitive impairment.

As discussed throughout this book, eating fat can be a good thing if we are consuming healthy fats. Healthy Fats provides the body with energy and lubrication for the brain and insulation for body organs and are essential for the absorption of fat-soluble nutrients (vitamins A, D, E and K). The three main types of healthy fats are saturated, monounsaturated, and polyunsaturated. Monounsaturated fats from olives and avocado assists with digestive and gallbladder health due to the chlorophyll and can be consumed as lamb, beef, and wildlife animals. Polyunsaturated fats can become toxic when used as cooking oil for frying, etc. and can increase inflammation, pain and depression risk; as a result, these should be eaten through nuts, seeds, leafy greens and fatty fish. Fats from animals, vegetables, nuts, and seeds that are extracted via a "cold process" should be integrated into daily schedule if possible. Eggs are rich in Choline and should be consumed daily for brain and memory health. Males often require more EPA, and females often require more DHA in their diets, but it is important to remember that DHA and EPA needs change through the life stages. DHA is often required at younger ages (for structural support) and EPA in later ages. DHA supplementation may assist younger people with managing learning and behavioural difficulties, and it is useful to know that the adult brain is approximately 60% DHA fat. Saturated fats have various benefits including anti-bacterial, anti-fungal and anti-inflammatory, and can be consumed through butter, coconut, tallow, suet (from cows and lambs), ducks, geese, chickens, turkey, and pig lard. The different types of saturated fats include Butyric Acid (found in butter), Lauric Acid (found in coconut oil and palm oil), Myristic Acid (found in dairy products), Palmitic Acid (found in meat and palm oil) and Stearic Acid (found in meat and cocoa butter).

Trans-fatty acids are known as unhealthy fats and should be avoided where possible. Remove Trans-fatty acids such as Hydrogenated Vegetable oils found in margarine, etc as they increase the risk for – Liver damage, reproductive health, respiratory health, digestive disorders, neurological damage, unhealthy weight gain and impaired immune function.

Essential Fatty Acids (EFA) are essential and can only by obtained from foods with three types including Omega 3, 6 and Arachidonic Acid (AA). As discussed throughout, our genetic history/bioindividuality determines how a person metabolises DHA and EPA. DHA and EPA are important as they improve communication between synapses, improve dendritic spines on postsynaptic neurons, improve brain volume, assist with treatment of brain injury, assist in management of depression, are anti-inflammatory and may assist with eyesight. The fatty acid Omega 6 is known as Gamma Linoleic Acid (GLA) and can be consumed from Borage seed, Evening primrose seeds, Black currant seed, Hemp and Spirulina. It may assist with dermatological issues, menopause, PMS, Uterine cramps, PCOS and endometriosis.

Saturated fats such as Medium Chain Fatty Acids may improve cognition and memory and promotes neuronal communication by increasing dendrites and can be consumed in coconut and palm oils. There is also a special type of fat called Phospholipids which comprise neuron membranes and support communication between neurons, such as Spinach and Krill oil. A lack of phospholipids reportedly increases the risk of depression and schizophrenia.

Prebiotics and probiotics are now common practice and form an important part of our nutrition balance. Prebiotics are often referred to as soluble indigestible fibre which helps clean the colon, encourages health microbes to be released, improves satiation and assists the microbiome "garden" in the colon by assisting the probiotics and microbiota to flourish and lessen the harmful bacteria to propagate. Examples include raw and cooked onions, garlic, Jerusalem artichokes, leeks, asparagus, wheat, beans, bananas, agave, dandelion root and chicory root. Chia is also a useful prebiotic as it contains soluble and insoluble fibre and is rich in omega three and is best when soaked in water prior to consumption. Probiotics are also known as Psycho-biotics and there are 400-500 different kinds of healthy microbiota in our gut. Probiotics can be used to promote a healthy digestive system, prevent infections, prevent diarrhoea, lessen inflammation, improve immune health, assists in the production of Vitamins K, B, Lactic acid and Folate, assists in GABA and Serotonin production and lessen anxiety and stress responses in the stomach. Fermented foods promote intestinal and brain health including yoghurt and cheese with live cultures, kefir products, sauerkraut, kimchi and miso, kombucha, brewer's yeast, Yakult, micro-algae. It is important to note that yoghurt probiotic function is often only active for several days, not weeks as advertised.

Modbiotics may also be beneficial for the body for people who have a poorer gut microbiome. A modbiotic is a type of prebiotic that helps to modify and regulate the gut microbiome. As well as the many other health benefits of Turmeric, some research has suggested that it may also have a modbiotic effect by changing the composition of the gut and increasing the richness of bacterial species. There are several companies that sell products with the modbiotic approach in mind, but this book will not be focusing on modbiotics due to how new the area is at present.

One of the reasons we should eat the aforementioned foods is for the essential vitamins and minerals our body needs. Essential vitamins can be water soluble or fat soluble. Water soluble vitamins are not stored in body, so risk of overdosing is low; whereas fat soluble vitamins are stored in body, so the risk of overdosing is higher. Fat soluble vitamins include vitamins A, D, E and K and are absorbed into fats but also need fats for the body to absorbs these.

Vitamin A helps immune system to protect against viruses, assists the eyes, skin and lungs, assists in helpful gene expression and can assist in reproduction. There are at least four forms of Vitamin A including retinol, retinal, retinoic acid and retinyl esters. One of the side effects of insufficient Vitamin A is corneal and other eye related difficulties and degeneration. Vitamin A is mostly stored in the liver, and when there is an insufficient level of Vitamin A ingestion, it will be slowly released from the liver.

Vitamin E contain eight different fat-soluble antioxidants that help the respiratory system and mitochondria and lowers risk of cognitive decline and Alzheimer's disease. Vitamin E also assists by preventing the spread of free-radical reactions in the body. One of the eight forms of Vitamin E (a-tocopherol) is contained is the body's plasma. Insufficient Vitamin E longer term may lead to poorer pain management, and neurological, muscular and skeletal health.

Vitamin K assists the body by functioning as a coenzyme for biological reactions involved in blood coagulation and bone metabolism. There are two different types K1 and K2. Vitamin K1 assists with cognitive function in older adults and vitamin K2 is essential for nerve health in the brain.

Vitamin D Is not really a vitamin but a Neurohormone which helps the body to absorb calcium, magnesium, phosphate, iron, and zinc. The Vitamin D receptors in brain also increase serotonin production and improves mitochondrial functioning and can assist people with Depression, pain and immune problems who often have low levels of Vitamin D. Vitamin D also assist in bone health by assisting the bones in their absorption of calcium. Vitamin D is high in foods such as fatty fish and eggs.

Vitamin C is a water-soluble nutrient that acts as antioxidant. It may help the immune system, improve mood and cognition, and assists in the biosynthesis of carnitine, neurotransmitters, collagen, and other connective tissue components, and modulates the absorption, transport and storage of iron. It can be found in banana pulp, citrus, pulp, and many other food sources.

Vitamin B has many types which generally supports the management of blood glucose, support serotonin, and improves cognition and neurological health. Vitamin B1 (Thiamine) assists in the metabolism of carbohydrates and branched-chain amino acids. Vitamin B2 (Riboflavin) assists with various oxidation–reduction reactions in several metabolic pathways and in energy production. Vitamin B3 (Niacin and Niacinamide) may help as a mood stabiliser and sedative, assists with fatty acid synthesis, assist with arthritic pain, and sometimes are used in natural treatments of schizophrenia and alcohol misuse. Vitamin B5 (Pantothenic Acid) is a water-soluble vitamin that is essential for fatty acid synthesis and degradation, transfer of acetyl and acyl groups, and a multitude of other anabolic and catabolic processes. Vitamin B6 is comprised of six related compounds and supports Neurotransmitter function, assists with insomnia, hypertension, poorer cognitive functioning, PMS, and irritability, and may assist with the symptoms of depression and anxiety. It is rare, but insufficient levels of Vitamin B6 may worsen depression and increase cognitive confusion. Vitamin B7 (Biotin) promotes appropriate function of the nervous system and is essential for liver metabolism. Vitamin B8 (Inositol) may be effective for depression, panic attacks, agoraphobia, obsessive compulsive behaviour, bulimia, and binge eating and may also be effective for SSRI resistant people. Vitamin B9 can be from folate (natural form and healthier option) or from folic acid (synthetic form and less healthy). Some of the difficulties with folate arise from needing levels of the enzyme 5-MTHF to convert folate, and approximately 40% of western populations have difficulties converting folate as a result. Some research has suggested that people with bipolar, depression, schizophrenia, Autism and ADHD are more likely to have a deficiency of the 5-MTHF enzyme. Vitamin B12 may assist with the symptoms of depression, fatigue, anxiety, cognition and psychosis, and sufficient levels are essential for healthy blood and neurological function.

Choline is required by the body for the structural integrity of cell membranes and is involved in methyl metabolism, cholinergic neurotransmission, transmembrane signalling and lipid and cholesterol transport and metabolism. Choline (Cytidine-5-Diphosphate Choline) is an essential nutrient that breaks down into Phosphatidylcholine, then into Alpha-GPC. Alpha-GPC, allows the choline to assist in memory and cognitive functioning, may assist in recovery for people with TBI, may assist with treatment of some addictions such as Cocaine cravings and may reduce manic symptoms in bipolar. Choline can be consumed through beef, liver, eggs, fish, chicken, and milk. Insufficient levels of choline may lead to liver damage.

Essential minerals help the body "get stuff done" and helps various body systems. It includes macro-minerals and trace minerals. As the name suggests, macro-minerals are needed in larger amounts and include calcium, phosphorus, magnesium, sodium, potassium, chloride, and sulphur. There are many different forms of magnesium and different benefits. In general, magnesium assists with stress, anxiety, cognitive function, and mood regulation, improves serotonin synthesis and reduces inflammation and oxidative stress. Magnesium Citrate is best for digestive health, bowel health, leg cramps, sleep; Magnesium Malate is crucial for ATP production and may assist with fibromyalgia; Magnesium Glycinate is great for calming, and back pain relief; Magnesium Aspartate may assist rapid cycling bipolar management; Magnesium cream may assist in management of muscle cramps; and Magnesium L-threonate was developed to cross blood-brain barrier to improve short-term memory and overall cognitive functioning, assist depression management and may dampen traumatic memories (such as nightmares). Potassium is the main intracellular cation in the body and is required for normal cellular function and has two types Orotate and Citrate. Potassium is often lower in people with PTSD, heart problems, bulimia, and anorexia. Calcium plays a key role in bone health and is also involved in vascular, neuromuscular, and glandular functions. Phosphorus helps maintain a normal pH in the body and is involved in metabolic processes. Sodium and Chloride are often found together in food and are necessary to maintain extracellular fluid volume and plasma health.

Trace minerals are required but only in smaller amounts an include iron, manganese, copper, iodine, zinc, cobalt, fluoride, selenium, chromium, lithium orotate and rubidium. Chromium helps in glucose regulation, may assist with depression and mood regulation and may decreases carbohydrate cravings and binge eating. Manganese is involved in the formation of bone and in specific reactions related to amino acid, cholesterol, and carbohydrate metabolism. Iron is an essential component of several proteins in the body including enzymes, cytochromes, myoglobin and haemoglobin. Fluoride protects against dental cavities, can stimulate new bone formation and is essential for the health of teeth and bones. Lithium Orotate is often used for cognitive and mood disorders, memory loss, alcohol recovery and relapse prevention, and may also assist thorough its neuroprotective and anti-inflammatory properties. It has also been found to help hippocampal volume and Brain Derived Neurotrophic Factor (BDNF), lessens risk of Alzheimer's and dementia, and assist in sleep quality. Selenium is an antioxidant nutrient involved in the defence against oxidative stress and regulates thyroid hormone actions. Selenium may also elevate mood, reduce inflammation, and enhance immunity. Zinc is crucial for growth and development as it facilitates several enzyme related processes related to the metabolism of protein, carbohydrates, and fats. Zinc also helps form the structure of proteins and enzymes and is involved in the regulation of gene expression. Zinc may also assist in depression management, bulimia and purging and anorexia disorders and may also assist the immune system. Copper regulates cellular energy, assists in production and neurotransmission, may assist in prostate health, and assists with depression management. However, some research has found that copper can be negatively impacted by antacid medications. Iodine assists with fatigue management and may assists in thyroid health. Rubidium may assist with depression management, stimulate the dopamine, norepinephrine and epinephrine pathways may assist in nervous and stress symptoms and is synergistic with potassium. It primarily comes from red meat, but can also be found in Brazil nuts, pecans, sesame and sunflower seeds, potato skins, eggplant, mushrooms, cucumber and avocado.

Coenzyme Q10 is a vitamin-like substance that acts like a vitamin is fat soluble and is found in mitochondria. There are two types Ubiquinol and Ubiquinone. Ubiquinol accounts for 90% of coenzyme Q10 and is the most absorbable form. It may assist in treatment of bipolar, depression, fatigue, fibromyalgia and schizophrenia and supports the mitochondria.

Phosphatidylserine is a phospholipid which is a key component of the cell membrane and is essential for the transfer of biochemical messages into cells within the brain and CNS. It also assists in regulating cortisol levels, increasing acetylcholine in brain, may be used in management of chronic stress, depression, ADHD, PTSD and brain injury and is found in egg yolks and liver.

Bromelain is a proteolytic that inhibits the

cyclooxygenase enzyme, reduce swelling and can be the foundation of anti-inflammatory treatment including for pain and fibromyalgia.

Acetyl-L-Carnitine (ALCAR) is made from L-Carnitine in the body and is often used as it may assist with cognitive improvement, and it increases acetylcholine, increases nerve growth factor in brain, increases metabolism and cellular energy and decreases insulin resistance. It crosses the blood brain barrier and is extra important for vegetarians.

Amino acids are the precursors to neurotransmitters, and are affected by our diet (low protein, poor fats). Some professionals use Amino Acid Therapy as substitutions for psychotropics and prescribed psychotropics. The top seven neurotransmitter essential nutrients include Free Amino Acids, Probiotics, B-Complex, Magnesium, Theanine, Curcumin with black pepper, Tyrosine.

And to finish this subsection, I thought it would be a tasty idea to end with chocolate. Chocolate (no added sugar or sugar alternatives) such as cacao nibs. Natural, sugar free chocolate has anti-inflammatory compounds including Flavanols and antioxidants (epicatechin highest in chocolate), may assist the liver and gallbladder as a bitter herb/food, may support mitochondria, may be used as a prebiotic via the flavanols, increases endorphins such as Dopamine, may increase focus, is high in magnesium and may assist the respiratory system.

Mental Health and Nutrition

We know now that the brain and stomach are connected via what is known as the enteric nervous system and the Gut Brain Axis. What we eat directly impacts our mental health for better or worse and directly influences our main brain centres including the Hypothalamus-Pituitary-Adrenal axis (HPA), Hippocampus, Cerebral Cortex, Amygdala and Brain Stem.

One aspect of nutrition and mental health that has received more attention in recent years is over-nutrition. In the modern world where there are vitamin pills for almost anything, it is easier for people to have too much of a particular nutrient. This overloading of a particular nutrient can cause many health side effects including a worsening of mental health. Some research found that people with an overload of copper, methionine, folic acid, or iron are likely to deteriorate if they take supplements containing these nutrients. Whilst it is true that good food is a primary need for a healthy self, a person may benefit from seeking professional assistance before self-medicating with vitamins. The professional such as dietician can assist a person in carefully identifying the specific nutrient overloads and deficiencies possessed by the person.

Chronic low-grade inflammation is caused by poor nutrition, stress and other and leads to further stress, early ageing, proinflammatory cytokines that suppress natural serotonin and neurotransmitter function, cardiovascular disease, metabolic disease, neurodegenerative disorders, shortening of the telomeres gene, cancers, poor mental health such as depression. Neurotoxins may impact mood stability that may present as a mood disorder such as depression. Neurotoxins may be ingested by eating foods that contain dough conditioners, artificial seasonings, yeast extract, synthetic carrageenan, maltodextrin, hydrolysed vegetable protein, WPC, aspartate and aspartame. Aspartame has been researched and some studies have suggested that it can worsen depression and irritability and may increase risk and severity of migraines, oxidative stress, diabetes, seizures, blindness, obesity, neurological disorders, and is considered a carcinogen. Sugar has also been theorised as a risk factor for maintaining depression. To help manage mood disorders such as mild to moderate depression various foods, herbs, minerals, fats and other should be ingested. Some of these include turmeric (with added black pepper to increase bioavailability), curcumin, ginger, black or green tea, berries, vitamin D, vitamin E, omegas, Kimchee and other fermented foods (increase Brain Derived Neurotropic Factor to lessen depression and dementia risk), vitamins and minerals with L-Methylfolate, Omega 3 Oil, Gamma Linoleic Acid (GLA) from borage or evening primrose oil, Free amino acids, Probiotics, Vitamin D, Vitamin B6, Niacinamide, Methylcobalamin, Lithium Orotate, Magnesium, Tryptophan (precursor for 5-HTP), L-tyrosine, and glandular foods such as Adrenal Glandular, and Hypothalamus Glandular.

Nutrition can influence and be used to improve other areas of mental health other than depression. If a person is anxious carbohydrates and health fats combined may help to lower their anxiety and lessen the need for comfort or craving related eating behaviours. Foods that are rich in Vitamins and Minerals with L-Methylfolate, Omega 3 Oil, Gamma Linoleic Acid (GLA) from borage or evening primrose oil, Free amino acids, Probiotics, Vitamin D, Vitamin B6 and B12, Niacinamide, Magnesium L-threonate or glycinate, Choline foods (Alpha GP-

Choline precursor), Tryptophan (precursor for 5-HTP), Lithium Orotate and Taurine can be helpful. Some examples of these foods may include bananas, figs, fruit vinegars, walnuts, and almonds. Support the production of GABA through food has also been suggested to lessen the severity of anxiety.
Glandular foods have also been used in some cultures to manage mental health difficulties. Glandular animal meats can include Brain, Hypothalamus, Pituitary, Adrenal, Thymus, Liver, Pancreas, Heart, Thyroid, Lung. They can be used to assist treating or managing stress, supporting immune function, metabolism, fatigue recovery and weight loss, managing depression and substance misuse recovery, supporting digestion, sugar metabolism and fat digestion, managing immune function and some research suggests it may improve cognitive functioning. Glandular meats have been traditionally prepared as Tripe, Lung, Thymus Sweetbread, Heart, Liver, Tongue, Testicles, Kidney.

Tracking our food can be difficult, especially when we live busy lives. There are many apps and methods that a person can use to track their food, however, most of the ways people track their food, don't consider their mental health. A 'Food Mood Diary' such as the one developed by Dr. Leslie Korn, can help develop insight including identifying patterns of food and mood and highlighting any possible addictive and mental health patterns. Ideas such as the food mood diary can also help a person manage disordered eating by identifying patterns and possible triggers. In the modern world overeating and still being micronutrient deficient is a commonly found issue with many clients who have high stress lives such as trauma, event stress, relationship issues, and other stressors. The overlapping and also competing ideas of overeating, rigidity in calorie counting and other behaviours have, in some cases, increased the risk of developing eating disordered behaviours in westernised countries. Disordered eating includes diagnosed eating disorders but also includes the behaviours that impact a person's function, even if they don't have a formal diagnosis. Regarding nutrition and disordered eating, there may be foods and food profiles that can assist in managing these behaviours. Foods that are rich in Tryptophan (healthy fats and complex carbohydrates) help to metabolise serotonin, and help a person feel full, which can be helpful in managing disordered eating behaviours. Some examples of these foods may include Whey protein (as it can help satiate a person and lessen purging related behaviours, whilst also containing the nine essential amino acids), oats and nuts and seeds. Another disordered eating behaviour that isn't officially recognised by the DSM-5 is Orthorexia. I and many other well trained fitness professionals have seen first-hand when the fine line between healthy calorie/macro counting turns into disordered eating. Orthorexia is thought to include – fixated, rigid, or righteous eating (Veganism, Keto, Raw Food only diets, etc.) being extrapolated to other people if it works for the one person and may also include the person becoming addicted to the claims of recovery. It is thought it normally starts with positive intent such as wanting to be healthier, or because of chemical/allergic responses to foods, but then leads to the unhealth expression of these behaviours.

Circadian rhythm has been discussed elsewhere, but I will briefly revisit it here in the context of mental health and nutrition. The term Chrono-nutrition can be defined as the time of day and meal frequency and refers to how nutrition and other foods are used to regulate the body's circadian and metabolic rhythms. A large amount of research has found that the disruption of circadian rhythm is correlated and sometimes causated with, insomnia, mood disorders, PTSD and other complex trauma. A metabolically healthy person has the physiological ability to switch between macronutrient sources for energy needs and to not substantially lose physical and cognitive performance in the short-term (24hours or less) in the absence of food. The body's circadian and metabolic rhythm are affected by factors such as sleep, physical activity (or sometimes lack of) and nutrients in the blood stream (including blood sugar and fatty acids). Difficulties here can contribute to Sleep/Wake disorders. Sleep/Wake disorders can be defined as longer term difficulties with sleep patterns and can include Dyssomnias, Parasomnias, Insomnia, and Circadian rhythm disruption. Eating foods rich in Vitamins and Minerals with L-Methylfolate, Omega 3 Oil, Gamma Linoleic Acid (GLA) from borage or evening primrose oil, Free amino acids, Probiotics, Vitamin D, B12, Magnesium Threonate, Lithium Orotate, Choline foods (Alpha GP-Choline precursor), Melatonin and Tryptophan (precursor for 5-HTP) can assist in the management of sleep/wake disorders.

Another area of mental health that can be impacted by poor nutrition includes alcohol and other substance recovery. Alcohol and other substance use has many causes and serves a variety of purposes. One of the causes for maintaining high levels of alcohol use includes the pleasure and or focus a person may temporarily feel from the dopamine rush from alcohol. Again, nutrition can be one of the tools a person can implement to lessen the need for the alcohol. Eating foods rich in L-Tyrosine (to increase the amount of dopamine secretion), Vitamin C (ensuring enough water is consumed), L-Glutamine (increase to lessen the physiological cravings for alcohol), Lecithin (increase to assist with Inositol, Choline and B Vitamins, and metabolise fats out of the liver), Chromium (increase to manage mis-metabolism of carbohydrates and help control blood sugar levels), Magnesium, Antioxidants, L-Methylfolate, Omega 3 Oil, Gamma Linoleic Acid (GLA) from borage or evening primrose oil, free amino acids, Probiotics, Vitamin D, Thiamine, Choline, Liver Glandular, Adrenal Glandular, Hypothalamus Glandular, Niacinamide, Potassium, Lithium Orotate, Tryptophan (precursor for 5-HTP), Melatonin foods and Thiamine.

Micronutrients are the elements required in small quantities to sustain life, and these have also been linked to mental health. Some examples include Magnesium and Vitamin B that have been demonstrated to reduce symptoms of depression, and anxiety, dietary fibre, probiotics, and prebiotics directly influence the gut health which is where approximately 90% of the body's serotonin is produced. When people are lacking in micronutrients it can be tempting to use multivitamins as the first and quickest way to fix this deficit. However, more recent research has found that using multivitamins as the primary or only tool of increasing a person's micronutrient intake can be detrimental and, in some cases, harmful to the person's health. The phrase "eating a rainbow" is still the best advice for many people to follow, as most people will benefit from this approach. Micronutrients and hormones that assist in the body's circadian rhythm include vitamin B12, lithium and melatonin. Novak et. al., (2021) reviewed two of the most important micronutrients – Vitamin D and Omega Fatty Acids.

Vitamin D is used in over 1000 different biological

processes in the body. This means that there are many opportunities for a deficiency, such as people living in the northern hemisphere or extreme southern hemisphere. Some of the Vitamin D receptors can be found in the cortex, hippocampus and cerebellum, and are used to help regulate movement, memory and cognition. Vitamin D also helps to transport serotonin to the brain and improves calcium absorption (alongside vitamin K). Vitamin D is also associated with an increase in Brain Derived Neurotrophic Factor (BDNF) which is thought to be important for neuronal health and growth, Glial cell Derived Neurotrophic Factor (GDNF), and nerve growth factor which is thought to help maintain and improve the structures of the brain. As discussed earlier, individual needs vary, but some research suggests that 10-25mcg of Vitamin D per day is beneficial.

Omega Fatty Acids including Docosahexaenoic Acid (DHA) and Eicosatetraenoic Acid (EPA) are considered healthy fats, while trans fats are known as unhealthy fats. Low DHA and EPA has been correlated with many mental health conditions including Schizophrenia, Attention Deficit Hyperactivity Disorder (ADHD), Depression and Bipolar. The main benefits of sufficient DHA and EPA include the inhibition of cytokine synthesis (which can cause inflammation when there is too much), assist in the dopaminergic and serotoninergic systems (which most mental health medications target), assist with receptor functioning, assist in the manufacturing of more synapses, assist in the maintenance of cell and neuron structure integrity and assist with neuroplasticity. The recommended source for DHA and EPA includes seafood related oils such as fish, krill and green lipped mussels, and nuts and seeds. Recommended dosages range from one to ten grams daily.

Macronutrients include protein, carbohydrates, and fats. They are the basic building blocks for energy and also provide important nutrition to the brain for emotional, cognitive and relational functioning. Hydration is also important here. Protein provides the brain with amino acids that are crucial for cellular integrity and the form the basis of neurotransmitters (such as Tryptophan and Serotonin) that regulate mood. When a person doesn't consume enough protein it ca lead to brain fog and a depressed mood. Carbohydrates include simple carbohydrates (fruits, vegetables and sugars) and complex carbohydrates (whole grains, starchy vegetables and beans) and provide the brain and body with fuel. When a person consumes simple carbohydrates, they receive a temporary boost in mood, but this can then cause a crashing effect in the body. When a person consumes complex carbohydrates, there is a slow release of glucose into the bloodstream. When a person doesn't consume enough complex carbohydrates their front brain functioning can be negatively impacted (thinking, problem solving) as can their other brain centres relating to mood and stress management. Fats can be naturally occurring (saturated, monosaturated and polysaturated) and engineered (hydrogenated/trans-fats). The natural fats help to lubricate the brain and body organs, help with memory, and transport vital minerals and vitamins to the brain. Whilst, engineered fats may impair learning and memory, lead to weight gain and damage the body's essential processes. Hydration is very important for a healthy body. Dehydration has been found to alter the function and structure of the brain resulting in problems in cognition, attention, focus, emotional control and irritability. A person needs to drink at least 1.8-3ltrs a day depending on weight, activity level, temperature, etc., to avoid hyponatremia (low blood sodium levels) and to avoid hyperhydration. Despite the various "optimal ratios" that are taught to us in the fitness world, told to us by nutritionists, etc., there is no one rule fits all to this ratio. Genetic heritage is a large determining factor regarding what ratio is best suited to an individual. The individual's needs are determined by their genetic heritage, current location, activity level, medication conditions, emotional state and current physical health. Groups such as the National Academies of Science have reference tables with the minimum amounts of macronutrients a person needs based on age.

Useful Nutrition Information

This small section will include some nutrition related information I found useful that didn't neatly fit elsewhere.

Hypoglycaemia is also known as Low Blood Glucose. It occurs when the level of glucose in the blood drops below what is a healthy normal level for that person. Some research has suggested that by stabilising our glucose levels, we can directly impact or influence our mood. There are reportedly two types of hypoglycaemia, known as Primary Hypoglycaemia and Reactive Hypoglycaemia. Primary Hypoglycaemia occurs from the inadequate supply of carbohydrates, whereas Reactive Hypoglycaemia is the excessive release of insulin following a meal with high amounts of refined carbohydrates. Hypoglycaemia can be a risk factor for people withdrawing from addictions such as tobacco, nicotine, and other stimulant related addictions, as the refined carbohydrates often release a shorter-term chemical high that the person may be looking for. The healthiest way to manage Hypoglycaemia is through consuming protein, healthy fats, starchy and non-starchy vegetables, complex carbohydrates and eating regular smaller meals. Hypoglycaemia can cause irritability, anxiety, nervousness, craving refined sugars, panic, crying, fainting, motor weakness or poorer functioning, personality changes, headaches, visual disturbances, confusion and shakiness. It may also be misdiagnosed and mistreated (but also comorbid with) – anxiety, bipolar, irritability, insomnia, tantrums, hyperactivity and depression.

As stated throughout the book, everyone is different, leading to people having different sensitivities and allergies. Food allergies, sensitivities and subsequent need for adapted or special diets are varied with some people having minor sensitivities to one food, while other people can have life threatening responses to specific food compounds. Food allergies often occur immediately after ingesting food but can take up to several hours later. Risk factors for food allergies include genetics, chronic infections, poor quality or pro-inflammatory foods, nutritional deficits and chronic stress. Food sensitivities and intolerances are the most common diet induced inflammatory response and symptoms can include skin changes, digestive issues, respiratory issues, fibromyalgia, GERD, IBS, obesity, migraines, depression, insomnia and chronic fatigue syndrome. In the westernised countries, one the most common food sensitivity or intolerance is lactose, with up to 70% of the westernised population being lactose intolerant. If your pulse raises more than normal after eating a specific food, it may be a sign of a food sensitivity. One of the other most common food sensitivities or intolerances is gluten. Gliadin and Glutenin are two proteins in grans that may trigger an immune system response. Type one sensitivity is known as Coeliac and is an autoimmune response in the small intestines. Gluten intolerance may trigger Gluteomorphins to be released in the brain which may then trigger an opioid-like effect on the brain. Type two gluten sensitivity is known as Non-Coeliac Gluten Sensitivity (GS) and is an immune response that leads to digestive and neurological difficulty and often goes undiagnosed. Approximately 1 in 250 people in the USA have Coeliac disease and approximately 1 in 10 people in the USA have gluten sensitivity. Alternative foods to gluten rich foods include Basmati rice, potato, coconut, almond, buckwheat, sorghum, sweet potato, legumes and tapioca. If a person is sensitive to gluten, there is approximately a 50% chance of the person also being sensitive to casein (found in milk). Some studies completed in USA have found that despite individual difference there are some geographic cultural averages regarding lactose intolerance. Cultures where dairy animals are not native and therefore people didn't genetically be exposed to dairy historically leads to intolerances such as 80-100% of African and Asian peoples are Lactose or dairy intolerant, 70-80% of African Americans and Mexican native peoples are Lactose or dairy intolerant, 60-90% of Mediterranean and Jewish descended peoples are Lactose or dairy intolerant, and 1-5% of Northern Europeans are Lactose or dairy intolerant.

A common medical disorder known as Alzheimer's is now being investigated as type three diabetes. This is because the conditions that lead to type two diabetes are almost the same as the conditions that lead to Alzheimer's disease and dementia. As well as lifestyle and nutritional changes such as managing the amount of grains and refined sugars ingested, there are other supplementary behaviours that may assist in lowering the risk factor for Alzheimer's disease. These include the use of Hyperbaric Oxygen Therapy, and by lessening inflammation in the body by consuming more turmeric, hops, rosemary, gingko, vinpocetine and vitamin E. Childhood trauma, toxic accumulations (such as from mould spores), brain injury, and the APOE gene are other risk factors.

Nutrition can directly influence the health of our Mitochondria. Mitochondria are known as the cells power plants and are essential for mental health. Viruses, stress, immune issues, poor nutrition and poor mental health can cause mitochondria dysfunction. To assist the body in its absorption of nutrients, improve gas, bloating and GERD, consuming digestive enzymes are important. The key Digestive Enzymes – Proteolytic Enzymes (digests proteins and inflammation), Amylase (digests Carbohydrates), Lipase (digests fats), Cellulase (digests fibre), Maltase (converts complex sugars into glucose), Lactase (digests milk sugars such as lactose), Phytase (helps to produce B Vitamins) and Sucrase (digests most sugars).

This section has focused on nutrition rather than weight goals or an individual's fat percentage. This is because a person's weight is only one indicator of their overall self and health and in our modern world is often used in a negative context. A person's weight and weight fluctuation can be influenced by bone density, water retention, genetic response to stress, and other factors that may not be considered in a medical or sales environment. The Body Mass Index (BMI) is still used today in a medical setting for a base line only and should not be used as a true indicator of a person's health. Focusing on weight also increases the risk of poorer mental health, which can then create a cycle of weight related difficulties and poor mental health. A person with an average BMI may be unhealthier than a person with a larger BMI due to fats being stored around the organs. This highlights the importance of speaking with your GP (General Practitioner) as a starting point and then speaking with a dietician regarding specific nutrition advice if weight and food related health difficulties are a concern. The GP can be a great place to start, however, as GP's know a lot about many areas of the human body, they can't be expected to have a full understanding of human nutrition as well.

Psychoneuroimmunology Understanding

This section will only be brief as it has a lot of overlaps with psychology. I have however, included this section as the books and online training did discuss ideas that I felt were noteworthy enough to have its own section.

Psychoneuroimmunology (PNI) can be defined as the relationship between the brain, our thoughts and emotions and our immune system. It primarily focuses on the interaction of our conscious thoughts and emotions and how they can directly influence our neurology and immune system. The mind-body is not mystical or magical as it is measurable in terms of a chemical response that is self-activated. This idea was first investigated in the 1950s by various doctors trying to understand the mind's manifestation from the western scientific cell-based approach. An emotionally induced illness is not imaginary due to the mind-body connection. The PNI research was demonstrated in 1975 and has been studying the discussed interconnections ever since. It is noted that the predominate theory behind the PNI area of study is no longer accepted by modern day immunologists, however, a lot of the PNI research is still useful regarding self-improvement and self-healing.

The 'PNI Global Awareness' training information defines spirituality as the personal development that can be defined as a person who uses all of their internal resources effectively. I like this definition as it can apply to everyone regardless of any formal spiritual or religious backgrounds and can encourage self-responsibility of one's health.

Whilst PNI research has a strong focus on internal healing, it does not dismiss the organic nature of illnesses.

Some organic causes for illness include bacteria and germs, genetics, and accidental injury.

PNI research acknowledges that traditional practices, alternative medical practices and holistic practices are effective at alleviating the symptoms of many illnesses. However, PNI research suggests that these approaches may be missing important information and approaches. PNI research discusses that the releasing of emotional suppression is required to ensure that the associated physical counterpart of the emotional pain is also released, often through physical intervention such as massage, acupuncture and other approaches. It also notes that helping professionals should be aware that they may accidentally compromise a person's level of overall wellness if they work with the mind and body as separate. This is because it is the direct relationship between how and where emotions are physically manifesting in the body that can determine the likelihood of change success. This success is therefore dependent on individual improvement, external support regarding emotions and thoughts (friends and or professional) and physical manipulation of the body's negative emotional manifestation.

A strong focus in PNI research is known as Biochemical Perception. Biochemical perception highlights what a person can do to change the negative biochemical responses to stress. How a person perceives an event, directly impacts our immune system, thereby affecting our emotional and physical health. This provides an understanding of the importance of managing internal and external emotional conflict. The biochemical perception is also about how our internal environment is affected by stress, and that our perception of this can influence whether the stress increases or decreases. Emotions have a direct impact on our immune system as the body communicates with the brain via bidirectional communication. Francisco Varela has previously defined the immune system as the body's brain, as the immune cells travel in the bloodstream throughout the body, they contact almost every other bodily cell. Every cell within the body responds to the way we think. The emotional aspects of a person influence their physical manifestation as evidenced by how stress modulates the activities of the nervous, endocrine, and immune systems.

The physiology of hopefulness can help the body to fight disease, whereas longer term emotional stress quietly harms the immune system and other system in the body. Nerve proteins known as neuropeptides affect our emotions as well as our physiology. A feeling in the mind will translate as a peptide being released somewhere in the body, as peptides regulate every aspect of the body including digestion and immune system responses. Neuropeptides' neuronal signalling of molecules influence the activity of the brain in specific ways.

Different neuropeptides are involved in a wide range of brain functions, including analgesia, reward, food intake, metabolism, reproduction, social behaviours, learning and memory. When we change the way we think, we can change the way we feel and this creates a perceptual change, which in turn allows for a deeper level of self-governance around our thinking. Every emotion has connection with a physical counterpart, and every ailment has an emotional attachment.

A researcher and author Candance Pert discussed that the body and mind are one and that what a person thinks and how they speak, directly impacts the state of the body's cells. The spleen, every lymph node, and all floating immune cells are in close communication with the brain. Emotions live and run every system of the body, and are in bidirectional communication with the nervous, endocrine and immune systems. Our emotional state is directly affected by perception, suggesting that negative mind will lead to an unhealthier body.

Our perception changes, life changes, etc. trigger an epigenetic response, meaning they change our genetic memory and gene expression. Genetic memory can be altered through the process of positive thinking and visualization (for those without Aphantasia), which stimulates a positive emotional state and adjusts to a negative one. Generating pleasant feelings helps a person to gain a sense of control, build hope, and increase the ability to generate and experience love, laughter, empathy, joy, confidence etc. The body's goal is homeostasis; however, the responses are not always accurate, and a stress response may be activated when there is nothing to fear, dependent on the emotions and thoughts a person is experiencing in the moments prior, during and following. Whilst the goal is to generate positive emotions and thoughts, to assist the body's health and recovery, we also don't want to suppress the negative emotions and emotional awareness. If we ignore our needs and suppress our emotions, our subconscious mind will alert us to the fact that something is wrong, which may result in a physical manifestation of emotional strain and medical illnesses. This highlights the ever-increasing importance of self-interventions.

Self-care is an umbrella term in this section will be focusing on what the individual can do to improve or maintain their health. However, it is important to remember that most of us don't live in silos and that the environment and people around us may either contribute to unhealthy levels of stress or be of assistance in managing unhealthy levels of stress. Self-care and self-efficacy can include choosing to be supported while remaining independent, managing our own perception of life, building emotional resilience, learning stress management techniques, learning relaxation, meditation and mindfulness techniques, promoting hope and happiness, proactively managing our internal emotional environment, understanding the power of our consciousness and taking appropriate levels of responsibility for the influence we have had and continue having on our health.

These ideas are useful to generate longer term health that in turn promotes natural chemicals in the body to assist in stress management. To further assist with self-efficacy, affirmations have been found to be a powerful wellness tool. Their aim is not to override the mind-body connections, but to relax, visualise (if possible), and generate within oneself an environment that is conducive to wellness. If the belief is health, safe and doesn't impact the person financially, the placebo effect can be of great benefit regarding some of these ideas, as for some people, the placebo effect allows them to build internal awareness of their needs. As stated throughout the book, poor or inadequate nutrition, a lack of emotional support, financial challenges and a lack of hope or purpose in life are all risk factors and, in some cases, causes for illness. Self-care and self-responsibility are important aspects of illness recovery. Self-care can be smaller things such as a longer shower, large things such as a holiday, and everything in between that assists a person to gain some mental and physical health benefit. Self-responsibility is taking an appropriate level of responsibility for any negative and positive outcomes that the person had control over. Assertive communication is another overlooked part of self-care. It allows a person to feel heard, connected and valued and therefore can lessen the number of negative thoughts and emotions a person holds. It also allows for the person to set healthier interpersonal boundaries and say no to taking on too many tasks. Assertive communication also lessens the need for aggression as the assertive communicator will often have less feelings of powerlessness or whelm, and less need to use aggression to process their negative experiences. Self-care and self-responsibility don't have to mean doing more, it can include finding time for oneself to relax.

The psychoneuroimmunology information has a strong overlap with various sections of this book including the holistic areas of health, as well as a strong overlap with psychology and the tools that will be discussed in the psychology sections.

Psychology of Body Brain Connection (Stress, Trauma, Enteric Nervous System)

Psychology of Human Body Brain Connection Introduction

The theme of this book has been how interconnected the body is, and this chapter will continue this theme. The human body is ridiculously connected, making both healing and injury more complex. The area of psychology is considered mental health, but the brain is physical; thought and emotion changes produce physical changes in the body, and life experiences affect us physically. This chapter will attempt to summarise this idea of mental health being physical, and will look at the effects of stress, trauma, substance misuse, and poor nutrition on mental health. This chapter won't be specifically focusing on therapy frameworks for several reasons. Firstly, my summaries would not do the frameworks justice, and secondly there is too much overlap between the frameworks, despite what some passionate single framework practitioners may believe. Even though a lot of research suggests that neurodivergent people are more likely to experience trauma and stress related difficulties, I have not explored this as again, I feel I would be summarising too much and may miss the complexities.

Many health professionals understand mental health is just as physical as a broken arm but with several important differences: the physical changes often happen in the most complex organ in our body (the brain), we can't touch or readily observe the brain so it is harder for the individual and treating professional to understand (and even when we see the brain through FMRIs, MRIs and PET scans, there is still a lot of guess work), and the research into understanding the brain is still only in its infancy. We also know now that we store stress all over our body, with some of the main skeletomuscular related areas including the trapezius, sternocleidomastoids (SCM), temporomandibular joints (TMJ), head, chest, psoas, pectorals, scapula region, diaphragm, pelvis, and iliotibial band.

Some of the things we have strong evidence for, regarding mental health and it's physical nature include: ASD may be organic with recent research suggesting a 70-90% genetic component, Schizophrenia may be strongly linked to dopamine (as evidenced by schizophrenia medication and Parkinson's disease studies), brain injuries can change an entire personality of an individual, ASD, ADHD, Bipolar Disorder, Major Depression, and Schizophrenia have a strong genetic component (with at least 2 genes being found to be responsible for regulating the flow of calcium into neurons and impacting their efficiency), poor nutrition not only affects cognitive functioning and maturation but also affects endocrine health and maturation in young people, and nutrient imbalances such as Copper overload, Vitamin B-6 deficiency, Zinc deficiency, Methyl/folate imbalances, Oxidative stress overload, Amino acid imbalances are risk factors for developing antisocial personality disorder, clinical depression, anorexia, obsessive-compulsive disorder, and schizoaffective disorder.

As stated, trauma can influence our gene expression, and not only increases person's risk of developing mental and physical illnesses but becomes a genetic risk factor for the next four generations if the gene expression occurs prior to reproducing. Some of our genes are turned off or on at birth depending on who the carrier is (male through sperm or female through egg) for example eye colour, height, etc. Whilst there are approximately 80 known imprinted genes, the research is not clear as to which ones are specific to the father or mother and how the genes choose which to express or recode. This image describes this in basic detail. The complexity of our genetics, combined with professional no longer being allowed to do direct genetic experimentation on human trials (due to ethics and morals), means we can only discuss risk factors. The research from the Minnesota Twin study strongly suggested that genes played a very strong part in behaviour expression, psychopathology, substance abuse, divorce, and leadership when comparing maternal twins. Epigenetics explains why people such as maternal twins who have the same genetic start (99% the same) can look and behaviour differently.

Once we think of mental health as more than thoughts and behaviours, but physical changes or differences, it become easier to understand the importance and relationship between epigenetics and mental health. Research by Harvard university has supported this idea. They found that positive influences such as supportive relationships and opportunities for learning and negative influences such as environmental stress or toxins leave a unique epigenetic signature on the genes' methyl markers, and this can increase the chances of the person and their future generations developing mental illness or other medical illnesses. The importance of movement and or exercise is also becoming discussed more frequently in mental health settings. More mental health professionals are starting to understand the importance of looking at the whole body and not just the mind in a silo. For some people formal exercise is too difficult, and for these people the idea of moving more may be sufficient. A research group developed an acronym known as NEAT standing for Non-Exercise Activity Thermogenesis, meaning to move more.

A detailed example of mental health impacting our physical health is discussed in the 'Anatomy Trains' book (see appendix). 'All negative emotion', says Feldenkrais, 'is expressed as flexion'. Hunch of anger, the slump of depression, or the cringe of fear many times and in many different forms. They all involve flexion. Among the quadrupeds, as we have noted, only humans put all their most vulnerable parts literally 'up front' for all to see. Subtly or obviously, people protect those sensitive parts: a retraction in the groin, a tight belly, a pulled-in chest. It is natural enough that when they feel threatened, humans should return toward a younger (primary foetal curve) or more protected (quadrupedal) posture.

As well as flexion in the body as an emotion response, negative emotion regularly produces hyperextension of the upper neck, not flexion as seen when the mastoid process is brought closer to the pubic bone. This not only protects the organs along the front, but also retracts the neck into hyperextension, bringing the head forward and down. The total posture, then, of the startled person involves rigidity in the legs, plus trunk and arm flexion, coupled with upper neck hyperextension. The problem comes when the startled posture is maintained (due to prolonged stress or perceived threat), which humans are perfectly and repeatedly capable of doing over an extended period. This posture and its variants can affect nearly every human function negatively, though breathing in particular is restricted by shortening of the SFL. Whilst discussing posture it is important to remember that our posture is not static, rather it is a dynamic counterbalance that is in constant change. Understanding this, can be beneficial as it allows a person to frame their posture as changeable rather than feeling stuck in a poor posture triggered by negative emotional, behavioural and environmental experiences. Our posture, like our cognitive functioning and overall health is changeable.

Easy breathing depends on upward and outward movement of the ribs, as well as a reciprocal relationship between the pelvic and respiratory diaphragms. The shortened SFL pulls the head forward and down, requiring compensatory tightening in both the back and the front that restricts rib movement. Shortening in the groin, if the protective tightness proceeds beyond the rectus abdominis into the legs, throws off the balance between the respiratory and pelvic diaphragms, resulting in over-reliance on the front of the diaphragm for breathing. The real, original startle response is marked by an explosive exhale; the maintained startle response shows a decided postural tendency to be stuck on the exhale side of the breath cycle, which in turn can accompany a trip through depression. This highlights the importance of having the body and the mind assisted together, where possible, such as in somatic psychology (i.e., Paired Muscle Relaxation).

Environmental influences such as trauma as well as life choices and/or life opportunities directly impacts our health and functioning, including our own gene expression (epigenetics). These environmental influences can then have the ongoing side effect of influencing future generations (up to four generations) if the gene expression occurs prior to reproducing. This means trauma can genetically be passed on, and positive life choices can also genetically be passed on. If a mother has PTSD the trauma can be passed through epigenetics impacting birth weight, and other physical stress responses such as metabolising cortisol. If a father has PTSD the trauma can be passed through epigenetics increasing rate of depression in the child.

What we call stress is often our cortisol and adrenaline response, which can be expressed physically as discussed, and can even be smelled by animals with dogs/cats/horses having the most research to back this up. Stress ages our DNA, increases the epigenetic expression of harmful responses, weakens our immune system, (often causing or being strongly correlated with various medical difficulties such as Fibromyalgia, heart failure, mitochondrial dysfunction, etc.) and lessens our cognitive resources (people when stressed, often report feeling like they have lost their memory or intelligence). I will briefly discuss the stress response in this subchapter including what is known as the fight and flight response. Fight, Flight, Freeze, Fawn, Feint, Flag, Flop, Fidget are all variations of the stress response a person can experience. The research history behind the terms influence how they are explained and which ones a professional may use in a therapeutic setting.

The freeze response can be seen when a deer is caught in the headlights. It involves the orienting reflex, an inborn impulse to turn your sensory organs towards a source of stimulation. Here the goal is to "stop, look, and listen" to better understand the situation and to determine if there is a threat. Your pupils will dilate as you turn your head towards the sound or sights that sparked your interest or concern. Most importantly, freeze occurs in preparation for action and is short lived.

The flight and fight stress responses are maintained by the sympathetic nervous system. This process involves initial attempts to flee danger; however, if it is impossible to escape you will resort to the fight response. The sympathetic nervous system increases blood flow to the heart and muscles of the arms and legs accompanied by faster and deeper breathing.

Simultaneously, skin will grow cold, and digestion is inhibited.

The fright response is a stress or trauma response that we see when flight or fight do not restore safety or there is no escape. The fright response may take over with feelings of panic dizziness, nausea, light headedness, tingling, and numbing. According to Schauer & Elbert (2010), this stage is considered to have "dual autonomic activation" seen in abrupt and disjointed alternations between sympathetic and parasympathetic nervous system actions. It is in this stage that we see the initial symptoms of dissociation.

The flag response occurs if there is still no resolution of the threatening situation. The flag response is the collapse, helplessness, and despair that signals parasympathetic based nervous system shutdown and immobilization. Dissociative reactions dominate this phase. Voluntary movements including speech become more difficult, sounds become distant, vision blurs, and numbness prevails. The heart rate and blood pressure drop, sometimes rapidly, which in some cases leads to the sixth stage, "faint."

The faint or flop response appears to serve several purposes from an evolutionary and survival perspective. When the body succumbs to a horizontal position blood supply increases to the brain. Furthermore, fainting is connected to disgust; an emotional response which rejects toxic or poisonous material. According to Schauer & Elbert, experiencing or even witnessing horrific events such as forced physical or sexual violence can trigger vasovagal syncope (Vagus nerve dysregulation) which promotes nausea, loss of bowel control, vomiting, and fainting.

The fawn response is often associated with trauma that involved a lot of emotional and attachment trauma. The person may be overly people pleases, scared to say what they think (overly passive communicator), rarely talks about themselves, may use flattery to avoid conflict, empathic to a point of negative functioning for self, easily manipulated or exploited, and a strong focus on social standing or social acceptance.

Finally, a less severe response that still impacts functioning and occurs as stress or trauma response is known as the fidgeting response. Fidgeting is often associated with boredom. However, whilst there are other causes such as stimulation seeking or sensory seeking, fidgeting can also be a trauma response when used as a distracting or grounding tool. This is often overlooked due to the minimal impact on functioning.

Relating to the fight and flight responses, some people can develop phobias of internal body states, such as the phobia of the upper body feeling "too relaxed". This is because the internal body state can then be associated with previous stress or trauma, and the brain then sees the relaxed state as a potential threat state. This can have obvious repercussions for a person trying to overcome their trauma related difficulties and is something professionals must think about when assisting people.

Adverse Childhood Experiences (ACE) include any negative, often traumatic experiences during the childhood (such as emotional, sexual, physical abuse, neglect) and can include discrimination. As the brain's right hemisphere develops first, any trauma received during young childhood impacts many areas of a person's mental health including increasing or suppressing negative emotions, withdrawal behaviours, difficulty with interpersonal and internal emotional attunement, difficulties regulating prosody (tone of voice), unhealthy neural circuits of attachment, poorer overall awareness of body, poorer self-regulation, difficulties experiencing or demonstrating empathy, and poorer intuition.

The trauma definition that I use for clients is "any event where a person feels helpless, hopeless or powerless may be catalogued by the brain as trauma", even if a person doesn't meet the DSM-V related criteria for a trauma diagnosis. This will often impact a person's biological, psychological and social functioning. There are many variations of health frameworks that are used by professionals and individuals. The fundamental framework I use is known as the Biopsychosocial model. The model includes the biological aspect of our health, the psychological aspect of our health and the social aspect of our health. Its simplicity allows for almost any person to learn it and expand on their understanding of their health. It also has room for added complexities when needed, such as expanding on one of the three areas in more detail such as social health. Many people forget that humans are animals, and we are a social species of animal, that greatly benefits from social interaction, and can have poorer health when our own individual social needs are not met (under or over social stimulation). This does mean that everyone needs to be an extrovert, rather it is helpful for people to understand their own social needs and aim to have these needs met. Research has suggested that healthy social engagement and social bonding can lead to internal feelings of emotional safety and emotional bonding and lead to healthy external behaviours of interpersonal contact and healthy proximity.

What we call emotions are neural and chemical states and pathways that are always changing. Emotions influence and are influenced by our perceptions, interpretation of internal and external stimuli, content of our memories, how our memories are encoded and retrieved, are attentional capacity and other executive functioning, our motivations, our gut health and many more. Our perception of reality dictates how we interpret (not see) the world around us. This perception is largely influenced by our life history, then the emotion at the time of the event/thing, then the context of the event of thing, then the event/thing itself, in real time, i.e., watching your favourite sports team or musician with someone who has an opposing view. As well as influence our daily idea of life, perception also influences personal and professional language around health such as blaming women, age, gender identity, etc. This is seen in the history and ongoing use of language when discussing mental illnesses and some physical illnesses. Medical and insurance-based terminology is still in practice across the globe today and often can lead to a negative perception of a person, sometimes even leading to the dehumanising of the person struggling. This may then lead to both the professional and the person struggling to have negative perceptions of self, leading to poorer mental and physical outcomes. By changing our language, we can change perception, and in some cases, this can lead to building or installing hope.

The importance of hope is often understated, but our brains are wired for change, and hope can encourage this neuroplasticity. This means we are all capable of thought and behaviour change, with a lot of work and a bit of good luck. This can be important for many people to know, as often when people are struggling, they can feel that this is the new normal rather than understanding that positive change is possible. Professionals can assist with this by changing language. When a person is dehumanised, hopeless, etc. they can then develop various unhelpful thinking styles. These often arise following trauma and can increase the stress responses in the body. Whilst there are many examples, some include catastrophising, overgeneralising and disqualifying the positive. If the person and or professional can help challenge these unhelpful thinking styles, it can lead a person towards hope, which may improve the outcome for the person struggling.

Regarding stress, there are some useful brain-change related information. Firstly, our hippocampus forms at ages two to three, which is why our earliest memory is rarely before two years of age. The more emotion, the better the memory's accuracy, and negative emotion can be more powerful than positive. However, too much negative/trauma emotion can lead to fuzzy memory. If there is too much emotion during an event, the stored memory can be fuzzy and then the person can inappropriately be triggered or have inappropriate generalisations and associations. I.e., all dogs are dangerous, instead of one specific dog is dangerous. The thalamus sorts incoming sights and sounds and then signals the appropriate parts of the cortex. This raw information is then interpreted by the cortex and an assessment of threat is made. If threat is relevant, the amygdala will then trigger the fear response, and emotional significance is added. Finally, the BNST (Bed Nucleus of the Stria Terminalis) perpetuates the fear response causing longer term unease regarding the stimulus as a safety response. However, this safety response if left on for too long, can turn into anxiety, trauma responses or other health difficulties.

Towards the end of this chapter various tools that people can self-practice, will be discussed. Many of these tools have modernised names and western scientific research as evidence, but most of these tools are still variations of mindfulness and meditation. I feel it is important to remember this, and to also remember that much of the current mindfulness and meditation practices originated from Tibetan, Buddhist and other Eastern practices including Yoga, Tai Chi, Qi Gong and Martial Arts.

Stress and Trauma Brain and Body changes

Stress and the cortisol response is good in small doses, however an excessive release of cortisol due to long-term stress damages our nerves and triggers a worsening stress response. The presence of stress provides an indicator that we are struggling, the feeling of a loss of control which leads to ineffective thinking. Sustained stress is a warning sign that a person needs to change or adjust their thinking or environment. Stress can be helpful for motivation, as it can help improve our focus, speed, and overall output, in small doses. It is the accumulation of unproductive stress that has the greatest negative impact on our functioning. It can remain active when we are constantly thinking and feeling pressure and/or when we feel powerless/hopeless. Positive change is still moving into the unknown, so a level of conflicting emotions is present.

There are many physical and mental health impacts of prolonged cortisol and adrenaline resulting from stress and trauma. I will include many examples and then briefly discuss other impacts and information relating to stress. The physical, emotional and behavioural impacts of prolonged stress and trauma include: a shutting down of the frontal cortex and associated regions, anxiety, aphantasia (lack of cognitive visual ability), attachment trauma responses (ambivalent and disorganised attachment are most common in people who experienced childhood trauma), autoimmune disorders (rheumatoid arthritis, multiple sclerosis, lupus, thyroid issues, Latent Autoimmune Diabetes of Adulthood/LADA diabetes, fibromyalgia), breathing pattern disorders, bronchial dilation, cardiovascular diseases, chronic inflammatory responses, chronic pain, contracting rectum, darkness on the eyelids, depression, diarrhoea or constipation, digestive disorders (Gastroesophageal Reflux Disease/GERD, microbiome issues), dilated pupils, dissociative disorders (Pain, substance misuse, eating disorders), eating disorders, emotional distress and other physiological response, excess glutamate triggering cell death in the brain, faster more shallow breathing, feelings of whelm, fluid retention, focus difficulties, gastrointestinal issues, gritting of teeth (TMJ), high acidity in the body (which can lead to a general stiffening of the body's fascial system), increase in heart rate and blood pressure, increase in sweating or shivering, increased anger and aggression, increasing heart contractions, inhibition of digestion, inhibition of salivation, insomnia and other sleep disturbances, intermittent explosive disorder, Irritable Bowel Syndrome (IBS), jelly-like legs, loss of hair quantity and health, memory difficulties, mitochondrial illnesses (mitochondrial diabetes, chronic fatigue syndrome, fibromyalgia), nausea, pain, perception/attitude towards food, poorer nail health, poorer self-image, postural issues, premature orgasm ejaculation, puffiness under eyes, relaxing bladder leading to increased risk of peeing, rigid or overly tense muscles, self-harming behaviours and urges (NSSI), stationary hyperventilation (observed as excessive sighing), stimulating epinephrine and norepinephrine release to unhealthy levels, stimulating glucose releases (increasing) by the liver, substance misuse, trembling, type 2 diabetes, and weight changes (loss of muscle, holding of fat).

As a side not here regarding medical illness relating to stress, it has been suggested that up to 66% of people with Fibromyalgia, and 76% of people with Rheumatoid Arthritis have trauma in their history, as well as at least 50% of people with IBS have a complex childhood trauma history.

Complex trauma is also a major risk factor of Central

sensitisation. Central sensitisation is a condition of the nervous system that is associated with the development and maintenance of chronic pain. It has two main characteristics that both involve a heightened sensitivity to pain and the sensation of touch, and these are known as allodynia (person experiences pain with things that are normally not painful) and hyperalgesia (when a stimulus that is typically painful is perceived as more painful than it should).

The disordered breathing mentioned above, leads to increased muscular pain, skeletal displacement, hyperventilation syndrome (over breathing), brain hypoxia, depression and headaches, and an overactive SCM (jutting out). This can also lead to jaw tightness, poorer posture and pain in the back and spine.

Grief and loss is another effect that trauma can cause but is rarely discussed. If the trauma has impacted the person's functioning to such a point that they lose the ability to complete their previous level of normal behaviours, they may experience an emotional response known as grief and loss. This can then become a further risk factor for developing other mental illness as well as worsening or developing poorer self-image.

Sensory sensitivities can also arise following trauma. Trauma can trigger or worsen sensory sensitivities such as light, heat, movement etc. This can be a direct response where the nerves are in a constant heightened or stress response level or can be a specific paired association trauma response where the sensory sensitivities is only triggered when specific environmental and internal conditions are met.

Prolonged stress and subsequent prolonged exposure to cortisol in the body, can also become a risk factor for developing psychosis. If a person already has a genetic risk factor such as a family member with hallucinations, then prolonged exposure increases the risk of this genetic factor being expressed. The person may experience a single episode of psychosis and then never again have this experience, or the single episode may change into regular experiences of hallucinations or delusions. Several brain related studies have demonstrated that when a person is experiencing a hallucination, the same area of the brain is firing as if that hallucination is real. For example, a person experiencing a tactile hallucination, will have the same tactile region firing in their brain, therefore making it real to that person. This psychosis related response to prolonged stress can also become confused and intertwined with other physical, sensory and neurological pain and experiences. Trauma can cause this prolonged exposure to cortisol if the person is unable to process the negative event(s) and subsequent impacts, further highlighting the need for sufficient support.

Prolonged stress and trauma can impact our sleep, leading to insomnia and impacting our circadian rhythm. Symptoms of a disrupted circadian rhythm caused by stress include – adrenaline, premenstrual tension, cravings, headaches, SCM and trapezius muscle pain and spasms, alcohol intolerance, indigestion, vertigo, poor memory, palpitations, ligament, laxity, insomnia, less tolerance for environmental changes, more emotionally reactive, poor concentration, fatigue, confusion, depression, physical weakness, food allergies, autoimmune diseases, hives and sensitivity to sunlight. Insomnia also increases the risk of low blood sugar (hypoglycaemia). For those who may experience hypoglycaemia, eating a small snack (55-85gram) before sleep can help break the cycle of middle insomnia. Lastly, Circadian rhythm disruption can turn into a disorder which may then lead to bipolar and bulimia. ADHD people can also be greatly impacted by a disrupted circadian rhythm more than non-ADHD people, and young people with symptoms of ADHD can find their symptoms drastically lessen through correct and healthy chronotype behaviours. Some nutrients that can assist in resetting the circadian rhythm include Melatonin and vitamin B12 (Methylcobalamin or Hydroxocobalamin are good, NOT Cyanocobalamin). Melatonin may help to regulate the if the person is experiencing circadian rhythm difficulties and B12 may assist in regulating the cortisol peak. Other nutrients and behaviours that may assist in lessening the insomnia and circadian rhythm difficulties include a magnesium sulphate bath, Epsom salt bath, oral magnesium dependent on bowel tolerance, sleep hygiene, and balancing hormones through diet/natural supplements.

Self-Image is one aspect of trauma that can be overlooked by professionals. Most people who have experienced trauma, have lower self-image, have poorer interpersonal boundaries and rarely use assertive communication. This low self-image can lead to avoidance behaviours such as passive communication as well as physical avoidance or isolation. These are known as safety behaviours. These help a person to feel emotionally safe but can often lead to further difficulties later in life. They can be very effective and even encouraged in the short term to allow a person to complete tasks, however, when replied upon they can become problematic.

This low self-image can be reinforced by societal expectations and can lead to a person wearing a metaphorical mask. When a person has experienced trauma, and a lower self-image is the response the mask and need to fit in socially, may often be used as its own type of safety behaviour. Societal ideas and education provided operate in two directions at once. It suppresses every non-conformist tendency through penalties of withdrawal of support and simultaneously encourages the individual to form values that force them to overcome and discard spontaneous desires. This may lead to the mask forming and becoming more entrenched in the person's idea of self. The need for constant support from other people may become so great that most people spend the larger part of their lives fortifying their masks. Often enough the individual becomes so adjusted to their mask, and identifies with it so completely, that they no longer sense any individual drive or satisfaction. This can lead to unhealthy ideas of self, goals, and of others.

As well as safety behaviours, the passive communication may lead to a build-up of unprocessed negative emotions which can cause a person to become more angry and eventually aggressive. Anger is often a symptom emotion triggered by other emotions and may lead to aggression, but doesn't have to, with the right knowledge and skill development.

Finally, pain is physical and medical difficulty that can be triggered or worsened by prolonged stress or trauma. Stress leads to muscle tension, which leads to pain, which leads to stress. Pain is an inflammation response that can lead to and be caused by substance misuse. Often caused by injury, poor nutrition and trauma. It can be helpful to understand that pain is part physical, part mental. The physical part of pain is the physical damage done to the nerves, muscle, organ etc, and includes the inflammation response that then elicits a pain response. The mental side of pain includes our relationship with pain, how much attention we give the physical pain and our level of stress or calm.

The hippocampus is surrounded by cortisol receptors. Overtime cortisol can cause atrophy and destroy the hippocampus, impacting trauma responses and memory. Some sensory information such as smell can skip the hippocampus and is thought to be the most powerful memory/trigger as it goes straight to the amygdala.

The amygdala is also known as the body's fear brain or smoke alarm, which means when the smell reaches the amygdala, the corresponding memory may be stored inaccurately depending on what emotion was being experienced at the time of the smell. A person with hyperarousal can learn to ignore danger signals in the amygdala in an unhelpful way such police officers who are used to being in this state can then miss actual danger in their environment or when a car alarm continues to be active for more than three hours, at first, we keep checking it, but eventually will ignore it.

The Insula and Amygdala communicate perceived negatives and feeds this system of negative thoughts and emotions and can cause an addiction to negative experiences. Smells and touch information bypass the Thalamus and go straight to the amygdala. This is why smells can evoke stronger positive or negative memories or emotions than sights, or sounds, and is important to remember for professionals who assist in trauma management. Paired Association triggers hyper-reactivity through the Insula. Depression and some addictions involve a hypo-activation of Insula often, because of complex trauma. A poorer functioning insula may impact one of senses known as Interoception. Interoception is a person's internal body awareness. Most people will understand when they are hungry, understand roughly what emotion they are feeling etc. When a person has trauma from a young age, it is a common response for the Insula in our brain to shrink thereby lessening or limiting the person's ability to understand their own body and emotions. This can lead to poorer outcomes across most life areas including health, relationships, employment, etc., and can be a strong correlator for many medical conditions. It also can impact our ability to understand pain such knowing when pain is safe such as safe stretching, and when is not safe such as being punched. The poor internal awareness of a person's emotion can be a form of poor Interoception and is known as Alexithymia.

The last brain area discussed here is known as the Cingulate Cortex. It is the self-regulation Centre, part of the Cerebral Cortex and Limbic systems and assists behaviour and thought decisions. It is used to assist with cognitive dissonance, and error detection centre but can often be impacted by trauma leading to poorer decision making, more negative or extreme thinking and greater emotional intensity. When the Cingulate Cortex is overactive the error detection centre is overactive, and this may lead to OCD.

Enteric Nervous System (nutrition and mental health)

Due to the discussed stress responses in the body, people will often either eat more or eat less when stressed. When the body is chasing a chemical high, the person will often eat more food of a poorer quality. This assists in the person' low mood for a short time, but leads to further low mood, fatigue and increases the risk of developing negative food and eating behaviours. When the body's stress response draws blood away from the digestive system and sends it to the muscles, the person may eat less as their food related stress response. This is because the idea of food, or the chewing of food may trigger nausea, or they may have less appetite due to the slowed digestive system. Furthermore, during this stress response, some people may experience nausea when chewing food due to the salivation response telling the digestive system to get ready, but the digestive system having less available blood to assist in the process.

When the quality of food (based on health) is poorer for an extended period of time, it can lead to a poorer microbiome. A poor microbiome is a risk factor regarding a person's psychological ability to manage and process trauma events. This is because the stomach is our second brain and what we eat impacts our mental health and vice versa. Our stomach communicates to our main brain via our gut-brain axis (enteric nervous system) and via our central nervous system and we have over 100 million nerve cells in the stomach. As well as the quality of food, we need to eat enough calories for our body including our brain. Our front brain requires a lot of calories often gained from glucose being converted from carbs. If we don't consume enough calories for our body, then the front brain doesn't receive enough fuel to perform its higher duties. As well as this, if we don't if we don't consume enough calories for our body longer term, we can start to lose muscle which may lead to an increase in joint pain and inflammation in the body.

Examples of Tools (not to be used as prescription)

There are many skills and tools a person can learn to assist in the management of the impacts of stress, trauma, AOD misuse and poor nutrition. Earlier I noted that the majority of the tools discussed can be traced back to various practices including yoga, tai chi, Qi gong and martial arts. This section will discuss different tools and skills including tools that are considered cognitive tools, breathing related tools, somatic and movement related and various other types of tools.

Mindfulness and meditation are the most well-known tools associated with psychology today. Mindfulness has its origins in meditation but there are some differences. Meditation and mindfulness are tools that can assist with accumulating positive emotions and processing away negative emotions; however, they don't just happen. Mindfulness and meditation are used to help calm the body, letting thoughts pass without judgement. This can be done as self-practice or practiced initially, with a professional. We don't want to ignore the negative or disturbing, rather find ways and learn tools to process, often with the assistance of a trained professional. Moving from a negative projection of self to a more positive projection is of benefit and can be achieved by a range of mind interventions that respect the need for interlinking the body with emotional wellness. Emotions generate like emotions (fear generates fear, hope generates hope, etc.), so acknowledging the negative emotions and working towards generating the positive emotions. The mindfulness definition I teach clients is "paying attention, to one thing, in the present moment, without judgement". This simple but slightly vague definition allows it to be used in various contexts with various tools such as eating, showering, lying down, muscle contractions, etc. It can be used in formal settings or added on to something the person is already doing such as showering.

Mindful eating involves a person holding food such as a biscuit or raisin in their mouth. The idea of the exercise is to develop insight and improve the person's Interoception as well as having the shorter-term benefit of grounding the person. It can evoke various emotions and when done in a professional setting can be useful in exploring thoughts such as control, patience, and emotions such as frustration or relaxation.

One example of emotional mindfulness is known as RAINe (Recognise, Accept, Investigate, Non-identify and evaluate). The person takes a moment to identify an emotion they are experiencing, then goes through the other stages with the end goal often being a calmer state and improved insight. Another tool similar to this is known as the visual shape and colour activity. If its established that the person doesn't have aphantasia or alexithymia, then the professional and person proceeds to the first step. The person identifies and emotion they are experiencing, then they find where it is in their body, they attach a shape/colour, then move the shape/colour to one of their feet, next they notice how the start and finish locations feel, notice if the shape/colour has changed and finally they notice if the emotion has changed. This is often abstract for many people initially; however, many people have reported positive outcomes.

There are many other types of visual mindfulness, and these can include building a vision of success using the brain's principle that if imagination can be real; visualising the negative thing drifting away such as on a boat or in clouds; visually adding a comedy element to a negative trigger; or visualising a target either physical or metaphorical. Some of the visual mindfulness tools can overlap with meditation, which is often thought to be a deeper mindful tool. Research has suggested that meditation may assist by boosting the immune system, reducing stress, dealing with negative emotions, lowering blood pressure, reversing heart disease risk, lessening substance use, managing weight, managing eating disorders, and improving sports related performance. Mantras are another tool that can be used in meditation and as daily mindfulness practice. Mantras can be simple such as personal mantras (I am kind), professional mantras such as (being assertive and kind helps meet the targets) and many others.

Other cognitive tools that may assist a person in managing stress including learning new things (brain training), doing old things a new way (brain training), reframing and the locus of control. Learning new things can be exciting and rewarding. If the person can focus on the direction of the new change instead of focusing solely on the achievement at the end, the journey becomes part of the excitement. Doing boring tasks in a different way can also help people to calm, refocus, and have more energy such as brushing your teeth with the opposite hand while moon walking. The locus of control and reframing often work well together. Reframing is changing the words we use with self-talk, talking about tasks and between people. Reframing changes our perception of the event, feeling, goal which increases the chance of success. The locus of control is focusing on the aspects we have control over such as when we shower, what food we choose to buy, etc. Sometimes there are negative events or memories where a person feels they are stuck and can't change anything, and this is when the reframing becomes important. Instead of "I am stuck in a job I hate" it might be "while I'm looking for work elsewhere, I will spend my time practicing healthy boundaries". Cognitive-based mindfulness such as gratitude journals or mindful reframing, improves front brain activation used in reasoning and problem solving, and is often the most beneficial when practiced regularly in low to medium stress states.

There are many forms of breathing practice that can be used to help a person slow down, calm, relax and focus, or even just build body awareness. Diaphragm breathing (also known as belly breathing) is a type of breathing that doesn't just use the diaphragm but also uses the vagal nerve to lessen the physiological response to stress and can be practice by almost anyone. The three types of breathing I like to use for myself and for people I help include ultradian breathing, humming breath and straw breathing, as all three have a lot of scientific research to support the positive effects and have been practiced in various forms for thousands of years. Ultradian breathing makes use of the brain's natural ultradian rhythm (brain hemisphere dominance changing in 90–120-minute cycles, by shifting contralateral nostril dominance). This communication is bi-directional and can be manipulated by choice. By closing the left nostril, we can increase left hemisphere activation and vice versa. Humming breath is very easy to learn. Recent studies have found a strong correlation between humming and decreased anxiety as the humming reportedly does several things – Increases Nitric Oxide which is helpful for blood health, the Rhythm activates the Vagus Nerve and calms the brain, it slows the speed of our breathing which calms the physiological responses and as a result calms the Central Nervous System. Various yoga practices have utilised the humming breath among other sound-based practices as a mindfulness and medication tool. The recent research now supports the humming breath, and the majority of the research suggests variations of the 'Mmmmm' sound to be the most effective at stimulating the Vagus Nerve. The straw breathing is also very easy to learn and can involve an actual straw or pursed lips. Breath research found that the average person will have 8-20 breaths per minute, a stressed person will often have more than 20 per minute, and research suggests the healthiest range is 8-12 breaths per minute. By pursing the lips or exhaling through straw after inhaling through the nose, the breaths per minute may decrease allowing for slower more full breaths.

The final type of breathing I practice comes from Pilates. It involves slow inhales and exhales whilst activating the transverse abdominus (TA). This is helpful as it strengthens some of the core muscles, slows breathing and helps build awareness.

One area of psychology that has seen a recent resurgence is the area of somatic and movement related psychology. This often involves movement or touching along with other forms of mindfulness. Common examples of movement that can be used for mental and physical health include Tai Chi, Qi Gong, Pilates, Yoga, Walking and any exercise that is safe and the person enjoys. Some research has found that when a person is physically moving forwards, it can stimulate the dopamine response and has a larger release when the person is focused on a goal.

People store stress, distress, trauma, and other negative experiences in their body – this may be improved through 'Body Orientated Psychotherapy' (Light touch massage, deep touch massage, movement related exercises without touch, energetic massage, joint manipulation, fascial massage and manipulation, lymph and nervous system massage, acupuncture, etc.). The Language of the body is touch and movement (body narrative), language of mind is words. Other body work such as progressive muscle relaxation, proprioception related exercises, acupuncture and sensory grounding can all assist with calming the body, and lessening stress held in the body.

Basic examples of proprioception include using a stress ball, having your feet on a balance board while seated and walking. The use of a balance under the feet while seated, in a therapy of medical appointment can assist in calming the person and improving focus. The body may find it difficult to maintain higher levels of stress with slow and flowing movements, so the balance board can be effective. A stress ball under a heel while seated can allow a person's natural fidgeting behaviour of the leg shake to become more beneficial longer term. Often people will leg shake to get rid of energy (often from nervousness, or adrenaline). Whilst this is useful short term, the leg shake often increases in speed and can lead to an increase in stress. The stress ball allows the leg to shake and absorbs some of the energy being generated thereby lessening the loop of stress created by a speedy leg shake.

Body-based Mindfulness such as PMR or sensory grounding, improves regulation (less reactivity) of reptile and mammal brains (bottom-up). Progressive Muscle Relaxation (PMR) can be done in full (often called a body scan) and is a very effective tools to relax the body before sleep, or it can be taught and practiced with one body part (often the forearm or quad controlled flex), and this can be used for in the moment tension reduction and stress reduction. This is a very useful tool for people who find the more traditional types of mindfulness related practices too abstract, as it provides almost instant physical feedback, and is practical or 'hands-on' type activity. Whilst often used by children, rocking can be beneficial with high stress difficulties for adults as well. Rocking may increase sleep spindles which are associated with improvements in sleep and can generate touch and feelings of being cared for which is a basic evolutionary need. Self-massage such as massing the ears (Auricular Contact) may assist in in activating the Vagal nerve as assist in calming to the surface and known as a Vitality Point in some health paradigms. Deep Massage and gently pulling the ear lobes and outer ear may also assist in relaxing, and energising and may release tension in the TMJ.

There is a movement type of practice known as Tension and Trauma Release Exercises (TRE). The TRE are a series of exercises designed to mildly stretch and stress the leg and psoas muscles (attaches – femur, pelvis, lumbar spine). Tightness in psoas leads to increased pain in neck, shoulders and lower back. Chronic contraction of the psoas increases the body's arousal or fight and flight state, leading to an increase in stress. The TRE activate the Central Pattern Generators up the spine leading to the tremoring response. This response is used by the body to lessen the arousal levels at all three levels (reptile, mammal and front brain). One side effect of the TRE practice is neurological exhaustion. When a person practices this for the first time, or after a long time between practice, they may experience physical and mental exhaustion, due to the body's arousal state and release of tension. As a result, many people who self-practice TRE often do so before bed.

Sensory grounding is often taught as an observational tool, where the person identifies five things they see, four then can hear, etc. However, I use principles taught in the fitness world as well as the sensory and OT related workshops attended, to teach a more practice and somatic sensory grounding practice. We have anywhere from 8-50 senses depending on how they are defined. However, in psychology we often discuss our five basic senses with people, and these can be used to help us relax/distract/ground and are also how our memories (good and bad) can be stored. The person identifies what their favourite taste, touch, sight, sound and smell are, and then they try and work on having a practical tool for at least two of these answers that they can carry on their person. Examples include a trail mix for taste, soft scrunchie or wrist band for touch, favourite photo or video saved on their phone for sight, music playlist or podcast saved on their phone for sound and favourite cologne or essential oil for smell. This version of sensory grounding practice enables a person to have a tool that they can use in the moment without having to try and use their frontal brain regions when they are struggling.

One practice that some people practice, and that Dr. Leslie Korn explains in fantastic detail, is known as Skin Brushing. According to research skin brushing can assist a person's lymphatic health, skin health, immune system, lessen congestion from allergies, lessen dissociation, may assist in managing mild to moderate depression, anxiety and stress, can be a self-care activity and can be used as a mindfulness activity. There are various versions however, the research suggests that there are specific pathways that should be followed for maximum benefit, so check with a trained professional first.

The final somatic related tool is known as either Saccadic Eye Movements or Saccade Stacks. Saccadic Eye Movements have been practicing in various forms in several spiritual practices such as in Buddhist meditations. However, it was officially introduced in a type of psychological therapy in 1989, and the thumb variation that I use and teach, I learn from the AMN academy holistic health course. Saccade Stacks are a variation of bilateral stimulation and can be done using pens (which is what most of the videos on the internet will use in their demonstrations), but this can also be done using your own thumbs. The Saccade Stacks help our 2 brain hemispheres communicate via the corpus Callosum. Short term this can be used to improve a person's ability to access their emotional experiences and long term can strengthen a person's Corpus Callosum. Originally this used to be done by the professional tapping on a person's head, or having a person follow the professional's pen, but limitations such as the client not liking their head touched, were found. Saccade Stacks also has the positive side effect of improving physical performance in fitness related tasks via neurological firing prior to performing the exercise.

There are many other tools that people can self-practice or practice with a professional that don't fit neatly into one type. The first one discussed is known as Animal Assisted Therapy (AAT). AAT often involves a trained mental health professional and their trained animal companion interacting with a person or group to discuss and improve mental health. A lot of research has found that the presence of a trained animal in the room assists in rapport building and calming for many people. However, it is noted that it is the professional's responsibility to ensure any protentional people seeing them are aware of the animal's presence as some people have phobias and allergies towards specific people. The most common animals used in AAT include dogs, cats and horses (equine therapy).

Rapport is a tool of its own, that may be forgotten about in many medical settings. Rapport is simply described as people feeling emotionally safe. There are various psychological studies, biochemistry studies and even magnetic resonance studies that have examined rapport either directly or indirectly. Most studies found that when a person feels emotionally safe, they are more likely to be honest, have better cognitive functioning and retain more information. They are also more likely to have better healing or recovery outcomes.

I will only briefly list some of the more technological advanced tools that can assist a person's mental and physical health, which should be discussed with a trained professional before using to applying to use. These include Cranial Electrical Stimulation (CES) (using a microcurrent to stimulate specific brain areas and releasing neurotransmitters), Photobiomodulation (PBM) (using red or near-infrared (NIR) LED light directly to the brain trans-cranially and intranasally) to stimulate, heal, regenerate or protect tissues), Hyperbaric Oxygen Therapy (HBOT) (inhaling 100% oxygen in a hyperbaric chamber that is pressured) and Heart Rate Variability Training (training or encourage a higher variability in the heart rate, although this area is still being studied).

Bilateral Stimulation is a practice of helping our two brain hemispheres communicate via the corpus collosum and research has suggested by this doing this a person can have better insight into their own emotional experiences longer term as well as help them calm in the short term. This can be done with walking, through with the previously mentioned saccade stacks and with audio sounds known as audio Bilateral Stimulation (BLS). BLS is a very easy to use practice that can assist in insight development longer-term, but also assist shorter-term in lowering the level of arousal a person is experiencing. By wearing headphones and listening to a program or app that has a simple tone going from ear to ear, the person can lower their arousal level. BLS is also an area that moves nicely into the area of sound therapies.

Sound therapies are used to assist in activating or reactivating the parasympathetic nervous system. As well as BLS, other examples include vibrating music tables, binaural music/beats, tuning forks, humming, vowel sounds, singing, etc. Very similar in theory but different in practice to BLS is Binaural Auditory Beats, where each ear is presented with a different tone entering the ear simultaneously, the brain then combines/reorganises the two different sounds and reconciles the difference into one coherent sound which helps to relax the body, lowering state of arousal. Sounds are used to restore patterns of vibration in the body through the conscious lengthening of a sound by using breath and voice, often a vowel sound, and often used alongside other tools such as yoga or movement. Often sound related therapies are used for pain management, and managing insomnia, PTSD, stress, pre- and post-surgery and pain addiction. The use of music and formal music-based programs such as the 'Safe and Sound Protocol' can be used to assist the Vagal system and assist in longer-term improvement in arousal management.

We often use sound therapy incidentally when we speak to other people. Many languages including English are dependent on hearing the harmonics of the word which can be adversely impacted by middle ear difficulties and by background sound. Emotions are expressed with different harmonics. Practicing overt communication of positive emotions utilises the harmonic element of language to further enhance positive emotion and thereby can improve mental health. This understanding of sound, emotion and tone also helps us to understand the importance of interacting with other people. This interaction is known as social health and not only assists in lessening avoidance, processing emotions, building connections, etc., but can also directly assist people in managing urge related behaviours and disorders by allowing a person to "ride the urge" until the urge lessens or dissipates.

Another tool that many people use as part of their daily routine is formally known as hydrotherapy. Formal hydrotherapy often involves specially designed baths where the water temperature can be controlled, along with sounds and smells, to assist in relaxing a person. This can be somewhat emulated with relaxing showers at home. The use of water and temperature can assist in managing pain, improving illness recovery, boosting mood and assisting various body systems such as the digestive and immune systems. When water is heated, it can help drive blood to the surface, when cooled it can help stimulate and drive the blood deeper into the body, and the alternating temperature can assist in stimulating blood flow and help move oxygen around the body. This can also be used with hot and cold packs when hydro-related practice is not viable. People such as Wim Hoff and practice such as Tumu breathing have used the idea of temperature to assist in illness recovery and even emotional insight development. For people who have access to baths and enjoy a bath instead of showers, adding Magnesium Sulphate, or Apple cider vinegar in water can assist in illness or tension improvement.

Nutrition has already been discussed in some detail but is again relevant to mention. Ensuring healthy levels of nutrients, fatty acids and minerals assists the whole body in manage stress and trauma related arousal. A lot of research has continued to highlight the benefits of omega 3's. Along with nutrition healthy exposure to natural light, especially early natural light has been suggested to improve overall health and assist in the management of brain injuries such as TBIs.

As discussed, the circadian rhythm is a very important aspect of our health and an area known as sleep hygiene discussed several of the elements that can be used to improve quality and quantity of sleep. When a person is in bed for more than 30 minutes and are still unable to sleep, some people find benefit from journaling their thoughts, this can be a list, story, or verbatim what they are thinking and can help to tell the brain to stress less as it is recorded and may assist the person in accidently processing some of the emotional cause of the thoughts. Sometimes when a person is in bed for more than 30 minutes and can't sleep, it can because they have too much physical energy, from either stress or adrenaline, or from not doing enough movement during the day for their body. In these cases, getting up and walking around or performing other non-stimulating activities can help a person feel tired. The 30 minutes is only an average, and some studies suggest 20 minutes, either way, the short period of time awake if left unchanged, can quickly change into a person still being awake four hours later. Other sleep hygiene tips can include ensuring the room is dark enough, the bed comfortable enough, the room temperature being cool or warm enough, eating enough, (for some people, if they don't eat enough, their body will wake during the night, impacting the sleep rhythm), limiting stimulants (sugars, caffeine, energy drinks, trans-fatty foods, etc.) before bed, practicing a regular routine helps the brain understand when to increase or decrease melatonin production (can be as simple what time a person goes to bed and wakes, or it can be detailed i.e., teeth, shower, toilet, read for 'X' minutes), managing screen use (using night time or orange light filters to lessen the blue and white artificial light), limit any mentally stimulating tasks such as stimulating games, articles, videos, etc., and the use of showering to assist in routine and melatonin management (relaxing shower at night and refreshing or colder shower in morning).

Finally seeking professional assistance for mental health is of great benefit, as even psychologists see psychologists. Research has supported the idea that when a person looks into the eyes of another person and discuss difficult topics, over a short period of time, both people's brain wave become very similar, activated by mirror neurons which, in this context can impact a person's emotions and empathy. It has also been suggested that the emotion mirror neurons are stronger than motor mirror neurons. When a person's motor mirror neurons are activated our skin cells and pathways help us understand it is not happening to us. Emotion mirror neurons don't have a skin barrier or defence, meaning greater emotional brain-brain influence, often called empathy. This is often a positive except for when it leads to manipulation/vicarious trauma, etc. There are many, many other tools used by professionals in therapy that I have not included here on purpose. The majority of the information in this book can be used by a person to assist themselves, whereas many of the tools I have left out the book often can be done incorrectly leading to a worsening of the difficulties.

There are many other therapies and practices that a person may gain benefit from that are not discussed in this book. Some examples include Acupressure or acupuncture, Alexander technique, Art therapy, Chiropractic medicine, Dance therapy, Energy Medicine, Environmental medicine,

Feldenkrais method, Massage therapy, Reflexology, Reiki and Tibetan medicine. If a practice is safe, healthy and helpful, often it is okay to practice. If a person is unsure, they should seek the opinion of a qualified professional in the area they are wanting to explore.

Information to discuss with a medical or allied health professional

Medication can be beneficial for some people who are struggling with their mental health, whether for shorter-term use or lifelong use. However, like anything we ingest, there are potential side effects the person should discuss with their medical professional. Some side effects often not discussed included on warning labels include mitochondrial dysfunction via inhibiting the mitochondrial respiratory chain, longer term use may increase toxicity in the body and may lead to other mental and physical health side effects, carnitine deficiency leading to lower energy, lower ATP production, impact sleep and lower nutrition absorption. Some Benzodiazepines are unhelpful and contra-indicated for PTSD as it has been found to increase anxiety, sleep disturbance, nightmare and irritability in most people, they may exacerbate suppression of autonomic nervous system and reduces healing mobilisation through the sympathetic nervous system. Stimulant medications have also been found to have similar effects when prescribed to a person during the early stages of unresolved trauma. The use of melatonin in children and teenagers impacts healthy teenage brain development. Prescribed medication including Sedatives, Hypnotics including Benzodiazepines may impact REM sleep and therefore the sleep cycle and may also increase risk of dementia. Nonsteroidal Anti-Inflammatory Drugs (NSAIDs) effectiveness is influenced by the time of day the person takes them for pain; in morning they may work with the circadian rhythm and at night they may work against the circadian rhythm.

There are many helpful tests for people that may require GP, or Dietician referrals, however, these may be forgotten, too expensive or deemed not relevant by the treating professional. These include Pharmacogenomics, Vitamin D panel, Tissue Mineral Analysis, Cortisol Specific Saliva Test APOE Gene test, MTHFR Gene test, COMT Gene test, Neurotransmitter tests, Hormone panels, Mould tests, Food sensitivity tests, Digestive and gut panels, Vitamin Deficiency test, Metabolic Test (fast, slow or mixed oxidiser test), and Mould test.

Psychology of Sensory Systems

This section will discuss our body's senses mostly from a mental health related angle. It will briefly discuss two alternative models for looking at our senses and then discuss one of the more current scientific models for our body's sensory experiences.

There is one model that breaks down our sensory experiences into 33 different senses. The 33-sense model has some outdated information and doesn't include other sensory inputs but can still be of interest to people when discussing the body's sensory information. This model separates the proposed 33 senses into nine sensory sections. Vision includes light, colour, red, green and blue wave; Hearing its own category; Smell is noted to have more than 2000 receptor types (not detailed); Taste includes sweet, salt, sour, bitter and umami; Touch includes light touch and pressure; Pain includes cutaneous, somatic and visceral; Mechanoreception includes balance, rotational acceleration, linear acceleration, proprioception, kinaesthesis, muscle stretch via Golgi tendon organs and muscle stretch via muscle spindles; Temperature includes heat and old; and Interoceptors include blood pressure, arterial blood pressure, central venous blood pressure, head blood temperature, blood oxygen content, cerebrospinal fluid pH, plasma osmotic pressure (thirst), artery-vein blood glucose difference (hunger), lung inflation, bladder stretch and full stomach.

There is another radical model of the senses that looks at 50 possible senses, although their definition of a sense is different to the traditional model. Again, some of their model may not be accurate, but has a lot of very useful information regarding our body both internally and externally. Their model has four main categories including Radiation Sensitivities, Feeling Sensitivities, Chemical Sensitivities and Mental Sensitivities.

The radiation sensitivities include sense of light and sight (including polarized light), sense of seeing without eyes, sense of colour, sense of moods and identities attached to colour, sense of one's visibility or invisibility, sensitivity to invisible radiation, sense of temperature and temperature change, sense of season and electromagnetic sense and polarity (including the ability to generate current as in brain waves or other energies).

The feeling sensitivities include hearing (including resonance, vibration, sonar, ultrasonic frequencies), awareness of pressure, sensitivity to gravity, sense of excretion, feel (particularly touch) on the skin, sense of weight and balance, space or proximity sense, Coriolis sense (Earth's rotation), and the sense of motion (body movement sensations and sense of mobility).

The chemical sensitivities include smell with and beyond the nose, taste with and beyond the mouth, appetite and hunger for food water and air, food obtaining urges including hunting and killing, humidity sense including thirst, evaporation control, acumen to find water and hormonal sense such as pheromones and other chemical stimuli.

The mental sensitivities include pain (external and external), mental or spiritual distress, sense of fear (dread of injury, death, or attack), procreative urges (sex awareness, courting, love, mating, child rearing), sense of play (sport, humour, pleasure, laughter), sense of physical place (navigation senses, position of celestial bodies), sense of time, sense of electromagnetic fields, sense of weather changes, sense of emotional place, (community, belonging, support, trust, and thankfulness), sense of self, (friendship, companionship, and power), domineering and territorial sense, colonising sense inc. receptive awareness of one's fellow creatures, horticultural sense and ability to cultivate, language and articulation sense, (used to express feelings and convey information), sense of humility, appreciation, and ethics, senses of form and design, reasoning, (including memory, logic and science), sense of mind and consciousness, intuition or subconscious deduction, aesthetic sense (creativity, appreciation of music and beauty), psychic capacity, sense of biological/astral time, (awareness of past-present-future events), capacity to hypnotise other creatures, relaxation and sleep (dreaming, meditation, brain wave awareness), sense of pupation- cocoon building and metamorphosis, sense of excessive stress and capitulation, sense of survival by joining a more established organism and spiritual sense (conscience, sublime love, ecstasy, sin, profound sorrow and sacrifice).

The nine basic senses include taste, touch, sight, sound and smell; as well as Interoception, Vestibular, Proprioception and Exteroception. Taste, touch, sight, sound and smell are the five most well-known senses, and it is because of Aristotle that many people still believe that we only have five senses. Three of the five basic senses (sight, sound, and smell) are known as having double organs – two eyes, two ears and two nostrils, which is suggested to have been useful in the wild from an evolutionary perspective. The Vestibular sense is responsible for balance, postural control, eye movements and alertness, but too much vestibular input reaching the brain may make a person feel nauseous. Proprioception is our body's awareness in space and is also involved in balance and the grading of force and pressure. Exteroception is how environmental sensory information impacts our functioning and is often referred to in more detail when discussing sensory quadrants or sensory modulation.

Interoception can be explained as the body's internal body awareness. This internal body awareness can encompass hunger, fullness, thirst, pain, illness, body temperature, sleepiness, toileting needs, anger, anxiety, distractions, focus, calm, boredom, sadness, personal space needs, time perception, and many more. When a person has poor Interoception regarding their internal emotional experiences this can lead to Alexithymia. Alexithymia is when a person struggles to identify their own emotions and as a result will often miss the emotional and physical warning signs when they are stressed, struggling or other. This can then lead to disproportionate responses to smaller stressful events or stimuli. It is very common in people who have experienced childhood trauma (as the brain's safety response) and neurodivergent people. A simple way of explaining this is a person without Alexithymia experiences warning signs of their emotions, then physiological changes and finally behaviours; whereas people with alexithymia often only understand their emotion based on physical and behaviour change only. Whilst this has the minor benefit of not consciously stressing over smaller stressors, it drastically increases the risk factor for developing stress related mental and physical illnesses.

Another one of body's senses is known as Neuroception. It is the neural detection of safety or threat based on the unconscious bodily awareness. This then triggers reflexive bodily changes in psychological states that serve as neural platforms for specific domains of behaviour. This is very helpful when the threat is shorter-term but becomes an issue (often leading to illness) when a person cannot not return to the rest state.

When discussing our body's senses in a health and mental health setting it is often helpful to look at if a person experiences the senses within the average range of if they have hypo and hyper responses. People who have experienced trauma and neurodivergent people are more likely to have sensory difficulties. One way of investigating sensory experiences is known as the Sensory Quadrants. The sensory quadrants include low registration, sensation seeking, sensory sensitivity and sensory avoiding responses. The person will answer questions with a qualified psychologist or mental health OT who then assists the person understanding what their responses mean on a Likert-like scale of one to five ('much less than most people' to 'much more than most people'). Another way of exploring the level of sensory input a person experiences is known as Sensory Modulation. Dr. Ayres' sensory modulation looks at three sensory modulation types (Sensory Sensitive, Sensory Slow and Sensory Seeking). When a person's sense is sensitive it means their brain doesn't block enough of the sensory input and as a result, their brain is hyperresponsive, such as experiencing moderate sounds as intense sounds. When a person's sense is slow it means their brain blocks too much of the sensory input and as a result their brain is hyporesponsive, such as experiencing strong smells as neutral. Finally, when a person's sense leads to seeking behaviours, it means the person is seeking or craving that specific sensory input to relax or stimulate themselves. A person can be sensitive in one sense and slow in another. Organic (from birth) brain difference, medical difficulties, trauma responses, brain injuries and other, may cause a person's senses to be sensitive or slow or seeking. It can be useful for a person to understand their own sensory needs to make their life more comfortable and also to lessen the risk of developing a mental or physical illness or having inappropriate behavioural responses to sensory stimuli.

As discussed, impacts of sensory overload or sensitivity occurs when the brain becomes overloaded by sensory information from the environment. This can lead to an increased risk of anxiety, increased risk of social isolation and avoidance as a safety behaviour, poorer focus, poor cognitive ability, poorer emotional understanding and management, larger even inappropriate or disproportionate emotional responses to stimuli, avoidance of specific foods, and can lead to sensory anxiety. Sensory anxiety is when a person experiences the traditional physical, behavioural and emotional difficulties of anxiety, but the cause is sensory related. This can make it difficult for a person to understand why they may be struggling and can make it difficult for them to seek appropriate professional assistance. The person experiences anxiety due to the discussed sensory overload, impacting and metaphorically, shutting down their frontal regions and triggering larger emotional responses.

A person who is sensory slow can also have behavioural difficulties and risks. A person who is sensory slow regarding smell and taste may be less likely to know if a food or chemical that isn't labelled is still okay to use or consume, the person who is sensory slow for sound may play their music too loudly leaving to socially related difficulties, a person who is sensory slow for touch may apply inappropriate levels of force when picking up objects and a person who is sensory slow for sight may damage their eyes from overexposure to artificial light more easily than the average person.

A sensory sensitive and sensory slow experience may lead to sensory seeking behaviours. Sensory seeking is when the person seeks the specific sensory input to help them calm, focus, etc. The behaviours can be small in practice such as smelling own cologne but can lead to social and other difficulties. It may lead to odd or socially unacceptable seeking behaviours such as sniffing people, licking objects, playing music too loudly in quiet areas, etc.

Martial Arts (Philosophies focused)

Focus

Some of the first records of martial arts being formalised dates to approximately the 20th century BCE, where murals have been found in ancient Egypt depicting wrestling techniques. Many cultures across recorded human history have had the need to defend themselves. The changing needs, belief systems and cultural practices influenced what was taught and what was considered effective. In more recent times, many researchers and martial arts enthusiasts have attempted to list the known martial arts from around the world, and along with online training options, the world of martial arts have become more global and accessible. Some people include distance weapons training as a type of martial art, however, due to my own personal background, I will be focusing on systems of cultural and self-defence practice that either originated or have a strong focus on self-defence and close quarters weapons. As such guns, bows and other such longer distance weapons focused martial arts will not be included.

People often ask what the best martial art is and some trained martial artists debate about what the best style is. Instead of focusing on what the best style or best martial art is, it is often better to identify what the purpose of training is. Is the training for fitness, self-defence, sports & competing, for personal growth, for internal healing or purely as a social activity. One of my instructors once told me that there are only so many ways the body can move. We are all limited to this notion and to the body's kinetic chains. It is often the practitioner of a martial art, not the martial art itself, and the purpose they train for, that determines if the martial art or style is "good".

There are many self defence companies and self defence systems that don't consider or sell themselves as a martial art, which I have not included in this section. As their purpose is self-defence, they have a different focus to most martial arts. Many of these companies and systems have great ideas, are easy to learn and can be very effective. However, I enjoy the respect for other people and respect for self, that is often left out of the newer self defence companies and systems but is often taught in martial arts.

Specific blocks, attacks, and other moves have not been discussed in this section for various reasons including not wanting to weaken the complexities of the martial arts, and not wanting to give information that may be unsafe (out of context). Almost all martial arts have their own idea regarding breathing techniques and the use of breath in attack, defence and healing purposes. They will often also have their own ideas about the best warm up routine specific to the focus of the martial art. The breathing techniques and the specific warmups will again not be included, but it is important to highlight this.

Previously I may have been considered a martial artist, however, I have not trained often or intensely enough for many years, and as such now consider myself a martial arts nerd. This section will start with only the martial arts I have trained in face-to-face, and finally finish with the martial arts I have studied on my own for general interest and the martial arts I only trained in for a small amount of time/dabbled in. I will be focusing more on the idea and/or philosophies of the martial arts regarding distance, attack, and defence, etc., from my understanding and experience.

Itosu Shito Ryu Karate

Itosu Shito Ryu Karate – My first exposure to martial arts was Itosu Shito Ryu Karate at the age of twelve. When I learnt this style of karate, there were three main parts – Kata, Bunkai, and Kumite. It is reported that this style of Karate may have the most kata among the many Karate styles, because the founder of Shito Ryu Karate (Kenwa Mabuni) was greatly influenced by two legendary karate masters – Ankō Itosu and Kanryō Higaonna. Itosu kata employ more powerful, explosive and linear techniques with long stances, whilst Higaonna kata involve shorter fighting methods with more emphasis on circular movements and the use of both hard and soft techniques. Following Kenwa Mabuni's death in 1958, the style of karate fragmented with many schools having their own variations of the katas and Bunkai. The majority of the katas could also be taught using a partner activity known as Bunkai. Bunkai symbolises the process of breaking down the movements of a kata to understand its self-defence application and is often referred to as the essence of kata. The Itosu Shito Ryu Karate Kumite had aspects of self-defence with a strong sporting influence. The sporting influence can be seen by the larger stances, the almost gliding movement of the competitors and the importance of controlling your strikes. The most common strikes in this style of Karate include, various punches and kicks, and can include leg sweeps. An extra training drill that was added to the training was known as Tan Gan Ho. Tan Gan Ho was the innovation of Sensei Fujimoto Sadaharu, which was taught to Shihan Kelleher and then my Karate instructor. Its purpose was to teach eye training, foot work development and arm Kumite development. Some of my most treasured sporting and martial arts memories came from my time training in this style of Karate. The main philosophies I took from this style were control of self through breath, muscle control, and control of self even when using aggression.

Rhee Tae Kwon Do

Rhee Tae Kwon Do – my exposure to Rhee Tae Kwon Do came in my mid-teens and due to circumstances outside my control, I only had the pleasure of training this martial art for a short period. Rhee Tae Kwon Do originated in Australia during the 1960's and was founded by Chong Chul Rhee, and is still seen as one of the traditional styles of Tae Kwon Do, and currently has no affiliation with the ITF (International Taekwondo Federation) or the WTF (World Taekwondo Federation). Rhee Tae Kwon Do gently highlighted to me the importance of fitness in martial arts. The philosophies I was exposed to during my training had a stronger focus on the fun aspects of kicking and the different distances involved with the different kicks learnt. There was some focus on hand strikes, but I felt kicking was still the primary focus. Their self-defence revolved a lot around striking and the kicking distance both in attack and defence. There were also various forms/patterns that helped a person practice the strikes into sequences often associated with your grading rank. The main skill I learnt from this martial art was the useful application of various kicking attacks and defences using the legs and the relevant distances required.

Wing Chun Kung Fu

Wing Chun Kung Fu – my exposure to Wing Chun began in my late-teens into early adult hood. The instructor ensured a strong focus on self-defence whilst keeping the principles of Wing Chun in the foreground. Wing Chun originated during the Qing Dynasty in Southern China and was introduced to Australia during the 1880's but became more well-known during the 1960's and 1970's. During my training, I found that there was a strong focus on the importance of fitness and how our physical fitness can influence our ability to defend ourselves and others. I found the primary focus of Wing Chun was Directness and Efficiency practiced through linear and simultaneous attack and defence principles. One noticeable difference regarding strikes, is that modern Wing Chun does not include kicks above the waist, as these are understood to putting the practitioner into unnecessary risk. The Wing Chun punch is often referred to as chain or roll punching, is very quick and linear, which allows the practitioner to maintain control over their centre and influence their opponent's centre of attack and defence. Various individual activities were used to teach the basic striking principles, and partner and group activities were used to teach and then test the striking and distance principles whilst under physical pressure. Weapon defence was also regularly taught, where the person could refine the Wing Chun techniques under pressure against various blade, blunt and other weapons. Forms were also used to ensure the person understood the structures of attacks and defences and were often associated with your grading rank. The biggest difference I noticed between the Wing Chun forms and other styles, was Wing Chun form was practiced with minimal floor movement required. My favourite aspect of Wing Chun was the effectiveness of attack, whilst minimising damage to self as the defender. The use of simultaneous attack and defence and the use of subtle angles allowed for this.

BJC Muay Thai

BJC Muay Thai – as I trained in Muay Thai and Wing

Chun in the same time period, my exposure to Muay Thai also began in my late-teens into early adult hood. The instructor ensured a strong focus on self-defence whilst keeping the principles of Muay Thai in the foreground and exploring some of the sporting elements of Muay Thai. Muay Thai has its origins in a traditional martial art known as Muay Boran. During the 1930's some of the variations of Muay Boran began to become regulated into a sport now known as Muay Thai. Muay Thai is known as the "art of eight limbs" as it uses both hands, both legs, both knees and both elbows. Despite Muay Thai now being used primarily as a sporting martial art, it can still make for a very effective self-defence martial art, as I was fortunate enough to experience. Muay Thai primarily relies on strikes from three ranges – kicking, punching and the knee range, but does make use of several sweeps and occasionally throws. Fitness and body conditioning are very strong focuses in Muay Thai, with Muay Thai practitioners being renowned for their high levels of cardio, toughness and "pound for pound" fitness and ability. My favourite aspect of Muay Thai was the simplicity of the martial art, and how quickly a person could learn the basics, and still be considered effective in its use.

Zen Do Kai Karate

Zen Do Kai Karate – my exposure to Zen Do Kai began in mid-adult hood. The instructor again ensured a strong focus on self-defence whilst keeping the principles of Zen Do Kai in the foreground. Zen Do Kai Karate was invented by Bob Jones, with the first school opening in 1970, with a strong focus on assisting people working in the security industry. This is evident when the various attacking and defending principles are discussed and can be seen in the various kata that form part of the Zen Do Kai Curriculum. Various strikes, throws and basic grappling techniques and principles are taught, and training includes individual techniques as well as partnered and group activities where the techniques can be practiced and refined. Weapon defence was also regularly taught, where the person could refine the Zen Do Kai techniques under pressure against various blade, blunt and other weapons. My favourite aspect of Zen Do Kai was how well rounded the martial art was in both practicality and skills learnt.

Pressure Point Knowledge

Pressure Point Knowledge (Martial Arts focus) – The term Pressure Point means the application of pressure to the surface of the body that is sensitive, which can increase or decrease pain. When used to heal, pressure points are used to slowly relax or release tension from a given area. When used in self-defence, the use of the sensitive or Pressure Point causes pain in the specific and sometimes referred area of the body, which may cause a person to stop the attack or be physically manipulated. Pressure Point use in martial arts has a long and varied history. The majority of self-defence martial arts such as Wing Chun and Zen Do Kai Karate, make use of the vulnerable parts of the human body to gain advantage in attack and defence. Some of the most well-known schools of Pressure Point use in martial arts are Aikido, Hapkido, Ninjutsu, Kyusho Jutsu, Dim Mak, Ju Jutsu and many, many others. Many security and military systems also make use of this knowledge. The main barrier to Pressure Point use as self-defence is that it often requires pain tolerance to be low enough for the opposing person to respond, which can be overridden if a person has had training, is in a heightened state of aggression or under the influence of substances. Also, individual body difference can make the use of Pressure Points as a primary form of self-defence dangerous as some people have cartilage, fat and muscle differences that can further impede the effectiveness of its use. Pressure points can also be used in healing practices with some styles of martial arts such as Kyusho, Dim Mak, Kalaripayattu, Tai Chi and many others, using Pressure Points as a way of healing self or another person.

Dabbled

The sections of Dabbled and Self-Studied is more for myself, so I don't forget the useful lessons learnt in the training and study. It is strongly noted here that I have not done enough training in these styles for my opinion to have much weight, and like the rest of this book, should be seen as only my experience and interpretation of these styles.

I was fortunate enough to have limited experience in various other martial arts, despite not being able to continue the training for various external factors. My brief experience with Aikido and Daitō-ryu Aiki Jujutsu focused on how easily the human body can be manipulated with minimal force and both had a strong focus on manipulation of the joint. My time with Krav Maga allowed me to see a different cultural influence behind a martial art. The simplicity of Krav Maga's direct attacks and aggressive defence combined with its high focus on fitness was helpful in gaining further self-defence knowledge. The many vulnerable areas of the human body were targeted and again the fitness training allowed for the techniques to be used under pressure. My brief time training in Goju Ryu Karate allowed me to see how similar and different a traditional Okinawan Karate style could be with a strong focus on close range training including grabbing and trapping techniques, throws and strikes and some focus on joint locks.

My experience with Brazilian Jiu-Jutsu demonstrated how effective grappling on the ground can be, including the many grappling defences, however, I found this the most difficult style to train in due to my own limitations, namely my overactive mind. Finally, the ISR Matrix seminar I attended taught practical skills such as closing distance, redirection, and basic grappling and disarming tools.

Self-Studied

When it comes to learning martial arts, face to face training is recommended and is often the most accurate way of understanding and practicing a martial art. However, when money, time or access are barriers, self-learning can be a useful tool to help a person understand the fundamentals and the physical and mental philosophies of the martial art they want to train in. This is relevant to this section of the book as it will highlight the Martial Arts nerd aspect of myself, focusing on the martial arts I have studied from books and DVDs, but have no formal training in or have only dabbled in. I have included this section as a memory backup but also to highlight the differences and similarities of martial arts. There are thousands of martial arts, and there are great martial encyclopedias that list many of them, however, I will only include the ones I have had the time and motivation to learn about.

Jeet Kune Do

Jeet Kune Do was a name that Bruce Lee gave his eclectic martial arts philosophy. For simplicity, the word style can be used, however, it does not fit neatly when discussing Jeet Kune Do. One of Bruce Lee's martial arts related regrets was naming his martial philosophy, however, the need for a concrete way of explaining his philosophy led to its name. He found that once he named it Jeet Kune Do, many people used the traditional martial arts notion of style and began labelling ideas and moves and Jeet Kune Do specific. Jeet Kune Do's philosophies were influenced by Eclecticism, Zen Buddhism and Taoism. Its self-defence related philosophies were strongly influenced by Wing Chun, Kickboxing, Fencing, Jujutsu and other practices such as boxing and street fighting. The underlying principle that the practitioner practices is focusing on a concept-based fluid practitioner, and not being fixed or overly patterned. Jeet Kune Do utilises sparring as a tool to practice the techniques, and to practice the key intention behind the majority of its attacks and defences, which is interception. Interception of the attack or the attacker is a key principle of Jeet Kune Do. Jeet Kune Do also involves understanding and utilising non-verbal cues, simultaneous attack and defence, economy of motion and fluidity. The striking influences are evident in the use of kicking and punching as defences, and the strong influence that Wing Chun had on Jeet Kune Do's philosophies is also highlighted through ideas such as centreline, trapping, kicking to knees and the simultaneous attack and defence. My favourite observation about Jeet Kune Do is the use of defensive kicks to the knee joint, as it allows for an effective defence with a simple and quick movement, whilst also maintaining a safe distance for the defender. The lead hand idea in attack is also very interesting.

Judo

Judo originated in Japan and was created in 1882 by Jigorō Kanō. Its founder had a history of training with various Ju Jutsu schools before founding Judo. Some of the biggest differences between Ju Jutsu and Judo include a focus more on free sparring compared to kata, as well as its strong focus on stand-up grappling and throws and the removal of many strikes and weapons training. Because of the large amount of throwing, there is a strong focus earlier on, in how to fall safely. The majority of training is completed as partner drills to allow for effective practice of the attacks and defences. There are six categories of throws that include hand techniques (Te-waza), hip techniques (Koshi-waza), foot techniques (Ashi-waza), back sacrifice techniques (Masu-setemi -waza), side sacrifice techniques (Yoke-sutemi-waza) and counter techniques (Kaeshi-waza). The throws normally have three stages to them including Kuzushi (encouraging the opponent to lose balance), Tsukuri (turning into and then fitting into the throw) and Kake (execution and completion of the throw). Currently, it is mostly practiced as a sporting martial art with many throws utilising the Judo uniform known as a Judogi, although there are still some schools that may teach it as a self defence system. Many of the throws and grappling defences focus on softness defeating hardness and evading the opponent's attack to cause them to lose balance. My favourite observation of Judo was the use of sacrifice throws, as the idea of giving the opponent access to your back is often seen as a negative in other martial arts but is capitalised in Judo practice.

Ninjutsu

Ninjutsu has a history that originally in approximately 570CE. It has had many different iterations and had many different contexts where it has been used. Famously it was used by various Japanese Daimyos as an official and unofficial branch of their military for both civilian and military roles. For simplicity reasons, this section will be focusing on Ninjutsu, in its more modern form, of today. This section will not be going into too much detail about the rich history, nor will it discuss the more mythical side of ninjutsu practice, often seen depicted in anime style cartoons. Modern ninjutsu gained popularity in westernised countries in approximately the 1970's when Massaki Hatsumi founded the Bujinkan organisation. The Bujinkan organisation slowly moved away from the label of ninjutsu because of its history and misrepresentation. It focused more unarmed combat training, conditioning and balance, and also took ideas from samurai philosophies and practices. One Hatsumi's students, Tanemura began the Genbukan organisation in the 1980's and focused on teaching Taijutsu, Bikenjutsu, Keishinteki Kyoyo, Bō Jutsu, Yumi, Naginata, Yari, Jutte, Kusari-gama, and shuriken. There are now various martial arts schools who teach elements that incorporate the principles of ninjutsu: unarmed combat utilising strikes, pressure points, grappling and throws, weapons training and agility and balance training. My favourite observation of modern ninjutsu is the agility and balance aspect of the training, as this skill is useful in almost all areas of modern life.

Karate

There are many variations of karate today, however, Karate was originally developed in approximately the 1400s, with some people believing it was practiced sooner (during the 900s). Many people believe that Karate originated in Okinawa and whilst this is true, it is also often left out of the teachings, that Okinawa was under the Imperial Ming Dynasty's influence and control during the 1400s. Despite the many variations of modern Karate, they often share similar principles. One of the first principles is Kihon, meaning 'basics' and often includes strikes (punches, kicks, open hand attacks, elbows), grappling and throws and their defensive counter parts, and some styles also include weapons training. Kata is another principle that is also taught and is used to help an individual understand the correct techniques and develop strength and agility and is also sometimes taught as a partner drill known as Bunkai. Finally, Kumite (sparring) is used to help the practitioner understand and practice the various techniques and drills. Some styles have a heavy focus on physical conditioning and breathing techniques whilst other styles have a stronger focus on kata. Some styles of modern Karate are geared towards sporting competitions whilst others remain a self-defence style. My favourite observation of karate from both a theoretical and lived experience perspective is how eclectic many styles of karate are often sharing many ideas and philosophies from Jujutsu, and even some styles of Kung Fu.

Philippine Fighting Arts (Arnis/Kali/Escrima)

The Filipino stick fighting martial is known as Arnis, Kali or Escrima, depending on what dialect lineage the particular school references. As I learnt about the fighting art under the name Escrima, I will be using this name for this subsection. Escrima originated in the Philippines before the Spanish invaded during the 15th century. As the history of this martial prior to this date was mostly anecdotally passed on, it is difficult to know when it was first developed. Modern Escrima has a strong focus on small swords, knives and sticks, as these weapons (tools) are based on the cultural influence and daily needs of the working-class population in the Philippines. Escrima spread throughout the westernised countries from approximately the 1920s through Hawaii and then across USA. The martial art itself discusses how the weapons used and represented are extensions of the body, and there are two main styles known as the hard styles and the soft styles. The focus, how the weapons are used and the movements in the "soft styles" often represent using bladed weapons such as knives and swords; whereas the focus, how the weapons are used and the movements in the "hard styles" often represent the use of blunt weapons such as sticks. Escrima is often practiced with individual drills and move sets for both offence and defence, as well as partner related drills and sparring. Empty hand techniques are often taught alongside the various weapon techniques. The most common Escrima weapons taught in the formal martial art include Short Sticks (Dolo Dolo), Single Stick (Rattan Stick), Double Rattan Sticks, Long Stick (Staff), Single Dagger and Double Daggers. There are various other weapons used by people living in the Philippines, often not included in the formal training of Escrima. My favourite observations of Escrima are how relaxed the arms are when using a weapon and how the various offensive and defensive drills allow the practitioner to use everyday items in self-defence if needed.

Aikido

Like Karate, Kung Fu, Jujutsu and other martial arts, Aikido is more of an umbrella name than a specific style. There are now many different schools of aikido such as Shodokan Aikido, Yoshinkan, Renshinkai, Aikikai and Ki Aikido. Modern Aikido began during the early 20th century and had spiritual underpinnings of peace and harmony, and most practitioners acknowledge Aikido's connection to Jujutsu. The core principles of Aikido include entering distance (Irimi), Breath control (Kokyu-ho), the Triangular principle (Sankaku-ho) and Turning movements that redirect the opponent's attack momentum (Tenkan). There are minimal strikes in Aikido, instead the practitioner primarily learns throws and joint locks that can be utilised for offence and defence. Because of the large amount of throwing, there is a strong focus earlier on, in how to fall safely. The majority of training is completed as partner drills to allow for effective practice of the attacks and defences. There are some schools of Aikido that still practice weapons training including the Jo, Bokken and Tanto. Some schools of Aikido also place a large emphasis on the health of the defender and the attacker, which may be related to modern legal systems but also the spiritual history of Aikido. The practitioner may also learn breathing and other related trainings to improve their inner ki (also known as qi in Chinese practices). My favourite observation of Aikido was the use of the attackers own body weight in the redirection throws, whereby their own momentum and weight, influences how much damage is done to their body.

Jujutsu

Like Karate, Kung Fu, Aikido and other martial arts, Jujutsu is more of an umbrella name than a specific style. This subsection will not be including Brazilian Ju Jutsu or Sambo as they have been redefined enough, to be considered their own modern styles. Some martial artists still consider Judo, Hapkido and Aikido as offshoots of Jujutsu due to their somewhat shared lineages and ideas. One of the earliest records of Jujutsu originated approximately 794CE and combined ideas from Sumo wrestling and various other self-defence ideas of the time (although it was not known as Jujutsu at this time). Modern practice began approximately between the 15th and 17th century CE as a form of unarmed combat. Jujutsu in its various forms often includes strikes (including hand strikes and kicks), grappling, throws, joint locks, and redirection of the attacker's momentum. As in Judo, and Aikido, the majority of the practice is done as partner drills, to practice the defences and attacks. Jujutsu today is normally practiced as a self-defence martial or a sports specific martial art, depending on the individual school or style a person chooses to train with. My favourite observation of Jujutsu was how eclectic it appears to be as a martial art, using ideas I have been exposed to in Karate, Aiki-Jujutsu and Aikido.

Hapkido

Hapkido originated in Korea following Japanese colonialism in Korea in the 20th century. It was reportedly founded by a group of Korean nationals with the most famous being Choi Yong-Sool. Its training philosophy shares many ideas with Aikido and Jujutsu such as redirection and fluidity with joint locks, breathing techniques, throws and grappling, but also has more "harder style" emphasis with their strikes including hand strikes and kicks. Hapkido also focuses on weapons training. My favourite observation of Hapkido is seeing the combination of its Aikido and Jujutsu roots being combined with the Korean martial arts cultural influence being expressed through the powerful aerial kicks.

Krav Maga

Krav Maga was originally developed during the 1930s by Imi Lichtenfeld in response to the violence being perpetrated towards the Jewish population and was eventually formalised for the Israel Defence Forces. It was influenced by Aikido, Judo, Karate and Boxing. The main focus of Krav Maga is dealing what it deems to be real world scenarios efficiently as possible. Its principles include simultaneous defence and attack, physical aggression in defence, completion of defence, situational awareness, mental flexibility, targeting the body's most vulnerable points and intense stress testing/training. Krav Maga teaches strikes, takedowns, groundwork, choke and hold escapes and empty hand weapon defences. My favourite observation of Krav Maga was the simplicity of its eight-point defence (effectively protecting the centreline) in allowing the defender to stay as safe as possible, whilst also allowing for a counterattack.

Systema

Systema appeared during the 1990s following the collapse of the Soviet Union and became used by the Russian special forces. There are various schools who claim to be the original or most effective style, however, they mostly share similar principles and ways of training. The four main principles in Systema include breathing practices, relaxation, importance of posture and continuous movement. These principles are important as they form the foundation of Systema and allow for effective use of their techniques, some which are unique to Systema. Systema makes use of their philosophies instead of specific techniques. The use of continuous movements in strikes, grappling, and weapon defence means specific techniques are not practiced in the traditional sense. Systema often also includes other weapons training, psychology of defence ideas and drills, partner and group training drills for self-defence, and whole-body attack, defence, and manipulation. My favourite observation of Systema is the use of breathing and muscle relaxation in both defence (to allow to absorb damage) and in offense (punching with a relaxed body) that results in movements that are less patterned.

Martial arts can improve the mental and physical health of the individual, improve social skills and confidence, and has many other potential benefits. Often it is the individual's training goals, and the instructor of a martial arts style that determines or at least strongly influences the individual's success, and not necessarily the style of martial arts itself.

Fitness Exercises – Pilates, Gym, Calisthenics and Yoga

This section will focus on exercises I have used or taught, will briefly discuss programming ideas, and will include a visual for each. There are many variations of exercises, some exercises can share the same name, and some exercises are called different names dependent on their historic origins and who taught the exercise. As stated earlier, I will be drawing from my own experiences so there may be some discrepancy in the name of certain exercises. There may be some exercises, progressions, regressions, and variations I have forgotten but as stated at the beginning of the book, my memory is the main reason for writing this book. There are many variations I have left out of the book, mainly because there are too many to include, and with the right base knowledge, the human imagination could think of an almost unlimited number of variations that took advantage of planes of movement, equipment differences, speed, muscle control, etc. Everyone is different in their fitness goals, journey etc. and this section is a giant list of exercise ideas, and not exercise prescription. However, there may be many exercises in this section that may benefit you but check with your own GP or fitness provider if you are unsure.

For people who are unable to do formal exercise, increasing your NEAT may be beneficial and easier to achieve. Examples on how to increase your NEAT include parking further away from a shopping entrance, taking stairs, if possible, instead of escalators and elevators, mindful fidgeting, throwing a ball to your pet, etc.

One area that is often forgotten when people begin their fitness or movement goals, is the awareness of our body.

Tools such as a body scan can assist in this understanding of the body. There are examples of exercises found in Pilates, Tai Chi, Qi Gong, Yoga, Circus, AMN Academy and many others that assist in the development of body awareness. The ability to isolate or activate specific muscles or groups of muscles can greatly improve a person's posture, technique, and physical ability when they perform an exercise.

There are many different theories regarding warming up prior to exercise and regarding what stretches are best. Depending on the professional you talk to, there are approximately eight different types of stretches. However, the three basic ways to define a stretch include static stretching, dynamic stretching and mobility stretching. Static stretching is often used at the end of a workout to assist in recovery, dynamic stretching is often used at the start of a workout to assist in warming the body and muscle groups and mobility stretches are often used at the beginning of a workout to assist in joint health.

There are competing ideas about when to exercise. For every study that claims mornings are the best there is a competing study suggesting afternoon and night. To simplify this, exercising in the morning may assist in burning fat throughout the day, may energise a person through the endorphin reward response and may assist in focus and establishing a routine. For some people exercising in the morning is unrealistic as they have difficulty waking early enough before work or study, their body doesn't have enough fuel to burn for the exercise and some people feel more tired throughout the rest of the day when they exercise in the morning. For these people afternoon or even night-time training may be more beneficial. Training at night allows a person to consume enough calories throughout the day to assist in ensuring sufficient training fuel, sleeping soon after a workout may assist in muscle recovery. For some people, training in the afternoon or night is unhelpful as the exercise may "recharge" some people, and therefore negatively impact their sleep. As discussed in different ways throughout this book, everyone has different needs and different exercise related goals, so the best approach is trial and error. Whilst this approach might be difficult and vague, it is often the best way for the individual to find what works for them.

There are also different types of exercises with different and overlapping impacts, that can often work well together, and are often suggested/prescribed to be practiced together (during the same workout or during the same week). Aerobic exercises assist in improving cardiorespiratory (circulatory and respiratory systems) endurance and have various forms such as low intensity (walking), medium intensity (light jog) and high intensity (High Intensity Interval Training/HIIT). Aerobic exercises have also been found to improve mitochondrial capacity as well as improving frontal brain functioning. Anerobic exercises help with building and maintaining strength. The force of the muscle contractions such as lifting weights can assist in improving strength, improving self-image, and may assist people with more severe mental health difficulties by lessening the dissociation by helping the person to ground themselves in the present. Resistance exercises help with strength and involves the body's muscles working against a weight from body weight (calisthenics and martial arts), resistance bands and other moving related equipment. Stability exercises assist with core and balance (Pilates) and can assist with improving self-image.

Finally, there are energetic type exercises such as Tai Chi, Qi Gong and Yoga that can assist with various physical improvements, mental health improvements and assist people interested in the spiritual aspect of their health.

I was originally overambitious for this section and was going to be including photos for all the listed exercises. However, after starting this goal, I quickly realised my error. As a result, I have decided to include the various exercises lists and only include photos of my favourite exercises (at the time of writing this book) from each section.

Pilates Exercises

As with most exercise, it is important to understand what the goal of the exercise of session is. Once the goals are defined, the followings ideas can be helpful when programming a Pilates session – ensure exercises chosen can assist the goal, exercises chosen to assist in gently correcting or supporting any postural issues or imbalances, ensure correct Pilates breathing is practiced, be aware of injury restrictions and if possible, try to include exercises from the frontal, sagittal and transverse planes of motion. Awareness of the body is an important skill to develop and maintain in Pilates, so ensure this mental component of Pilates is including in the training.

List of Pilates Exercises

4 Point Bent 90-degree Leg Lift and Into Chest

4 Point Bent 90-degree Leg Lift Hold

4 Point Bent 90-degree Leg Lift Pulsing

4 Point Cat

4 Point Cow

4 Point Hamstring Curl

4 Point Hold – Knees in-line with hips, hips are square, wrists in-line with shoulders, activation of transverse abdominus and looking slightly forward.

4 Point into Arm Lift (3 Point)

4 Point into Opposite Arm and Leg Lift / Bird Dog - – Knees in-line with hips, hips are square, wrists in-line with shoulders, activation of transverse abdominus and looking slightly forward. Slowly lift the opposite arm and leg and hold, ensuring the hips are still square, knee on ground is still in-line with hips and wrist on ground is still in-line with shoulder.

4 Point Knee to Elbow

4 Point Knee to Elbow Variation Hydrant

4 Point Plank

4 Point Straight Leg Lift

4 Point Straight Leg Lift Circles with Heel

4 Point Straight Leg Lift Hold

4 Point Straight Leg Lift Pulse

4 Point Straight Leg Lift to Side

4 Point Straight Leg Toe Tap

4 Point Straight Leg Toe Tap and Slide in

4 Point Straight Leg Toe Tap and Slide to Side

4 Point Thread the Needle

4 Point Triceps Pushup

Adductor Stretch Seated Straight Legs

Back Extension Breaststroke

Back Extension Hold

Back Extension Swimming

Back Extension Variation Dart

Bilateral Femur Circles Supine

Bilateral Overhead Arms Supine on Foam Roller

Bilateral Shoulder Circles Supine on Foam Roller

Bilateral Shoulder Drops Supine on Foam Roller

Bridge

Bridge Variation Lift and Slide

Bridge Variation Single Bent Leg Lift

Bridge Variation Single Straight Leg Lift

Buddha Stretch

Buddha Stretch Supine – feet are hips width or slightly wider than hips width on ground. Spine is neutral. Bring back of hands together, slowly raise arms over face, open up arms, then make a circular motion back to the beginning.

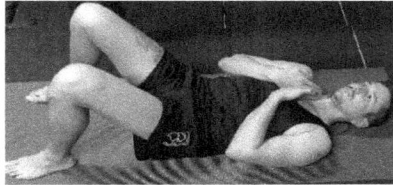

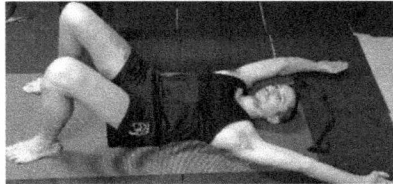

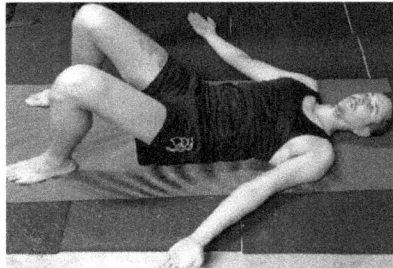

Butterfly Stretch Seated

Calf Raise

Calf Stretch Seated

Calf Stretch Standing

Calf Stretch Variation Tibia Wall Standing

Child's Pose Variation (1 Arm in front, 1 to the side underneath)

Child's Pose Variation (Arms behind)

Child's Pose Variation (Arms in front) – sitting with buttocks on heels. Either gently rest head on the ground and rest arms in front or stretch arms as far forward as possible to add stretch to the latissimus dorsi.

Child's Pose Variation (Lats with Arm Cross Overs)

Cossack Twist

Deadbug Hold Supine

Deadbug Variation Alternating Single Leg Drop

Deadbug Variation Opposite Arm and Leg Alternating – Neutral spine, hands in-line with shoulders, knees in-line with hips, creating an approximate 90-degree angle between the calf muscles and the upper leg. Slowly drop the opposite arm and leg whilst keeping correct posture as stated. Slowly alternate sides.

Double Torso Curl/Teaser

Double Torso Hold

Hamstring Stretch Supine

Hip Flexor Stretch in Supine

Hip Flexor Stretch in Supine 1 Leg

Hula Hoops / Trunk Rotations

Hundreds Supine

Hundreds Variation Bridge

Hundreds Variation Legs Straight and Shoulders off Ground

Hundreds Variation Legs Straight V

Hundreds Variation Tabletop

Hundreds Variation Tabletop Shoulders off Ground

Knees to Chest Relax

Knees to Chest Rocking

Knees to Chest Stretch Variation One Leg Straight

Leg Opening Supine

Leg Opening Supine Variation Heels on Wall

Lunge Stretch Variation

Lunge Stretch Variation Pelvic Lift

Lunge Stretch Variation Pelvic Lift Pulsing Forward

Mermaid Stretch

Mermaid Stretch Easy Variation

Neck Rolls Supine

Pec Stretch Bent Arm

Pec Stretch Straight Arm

Pilates Ring Chest Press

Pilates Ring Leg Raises

Pilates Ring Shoulder Press

Pilates Ring Side Position Pulses

Pilates Ring Squats

Pilates Ring Steering Wheel Front Hold

Pilates Ring Straight Leg Hundreds

Pilates Ring Bridge – Feet slightly wider than hips width, shoulders on ground, activating transverse abdominus, Pilates ring in-between legs, and gently squeeze ring with thighs and hold position.

Pilates Ring V-sit-up

Pilates Ring V-sit-up Hold

Plank

Prone Leg Stretch

Quad Stretch Standing

Quad Stretch Standing Variation Pelvic Lift

Rear Tabletop Hold Legs Straight

Rear Tabletop Variation Glute Pulses Legs Straight

Rear Tabletop Variation Leg Lift Legs Straight

Rear Tabletop Hold Legs Bent

Rear Tabletop Variation Glute Pulses Legs Bent

Rear Tabletop Variation Leg Lift Legs Bent

Scapula Dips

Scapula Pushups

Scapula Squeeze and Release Supine

Shoulder Circles Seated

Shoulder Circles Standing

Shoulder Circles Supine

Side Band Walk

Side Position Clam

Side Position Flying Clams

Side Position Inner Thigh Lift

Side Position Kick to Front

Side Position Side Kick

Side Position Straight Leg Heel Alphabet

Side Position Straight Leg Heel Circles

Side Position Straight Leg Hold – Hips in-line, bottom leg on the ground trying to have bottom foot in-line with hips, head either gently on bicep or gently in hand. Top leg up off ground, top foot slightly rotated to assist in gluteus Medius activation and hold top leg up.

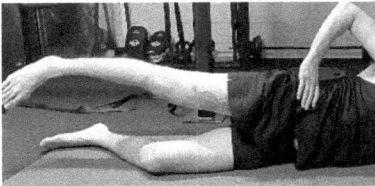

Side Position Straight Leg Pulse

Side Position Straight Leg Top Leg Hold, Bottom Leg Pulse

Side Position Straight Leg Double Pulse

Side/Oblique Knees Plank

Side/Oblique Plank Hand

Single Leg Circles

Single Leg Circles Variation Bottom Leg Straight

Spine Curls Supine

Spine Rolldown and Rollup Standing – Standing feet parallel, shoulder gently pulled back, chin gently touching chest. Slowly roll down as close to vertebra at a time as possible, until finger touch ground. Then slowly roll back to standing.

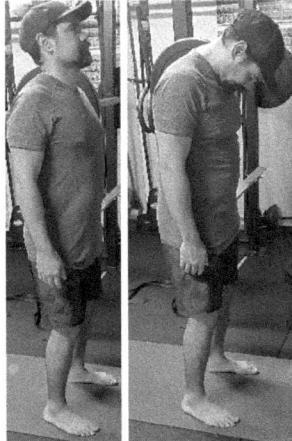

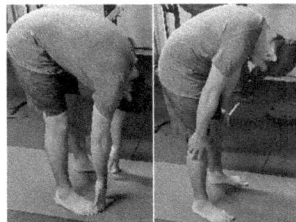

Spine Rolldown Variation into Plank and Rollup Standing

Spine Rolldown Variation into Pushup and Rollup Standing

Spine Rolldown Variation into Shin Hug and Rollup Standing

Spine Rollup and Rolldown Arms to Ceiling

Spine Rollup and Rolldown Assisted (Hands on Hamstrings)

Spine Rollup and Rolldown Wrist to Knees

Spine Rollup and Rolldown Weighted

Spiral Stretch Seated

Starfish Relaxation

Tabletop Hold Supine - Neutral spine, knees in-line with hips, creating an approximate 90-degree angle between the calf muscles and the upper leg, and Hold.

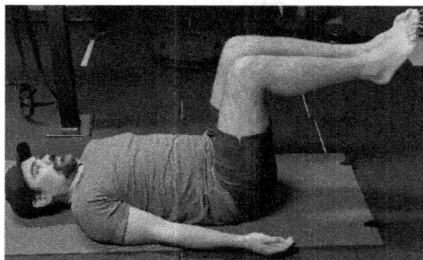

Tabletop Variation Dead-bug Alternating

Tabletop Variation Dead-bug Hold.

Tabletop Variation Straight Leg Circles

Tabletop Variation One Leg Drop to Toe Touch

Tabletop Variation One Leg Drop to Toe Touch and Slide

Tabletop Variation One Leg Straight

Tabletop Variation Touch Toes

Tabletop Variation Touch Opposite Outside Foot

Tabletop Variation Open-Close

Triceps Overhead Stretch

Wall Slide/Angels

Windmills

Windmills ½

Glute Stretch Combo Parts 1-3 – Neutral spine, bring the right ankle to the left leg's quad and hold. Then gently bring the left leg towards face keeping right leg in roughly the same position, and hold. Finally, bring left lg's heel to the right glute and hold. This sequence can be done with the head on the ground, or it can be done with head off the ground, slightly sitting up.

Glute Stretch and Twist

Gym

Writing a training program is difficult and writing one for use at a gym can sometimes be more difficult. Some people prefer to train their whole body, while others prefer to train specific muscle groups. As with all programming, identifying the goal is the most important first step. After this identifying fitness level baselines is important and can include how many minutes per kilometre, how many seconds per 100 metres, one rep max weight, many number of reps, etc., ensuring good technique throughout. As a very basic idea, if the goal is endurance smaller weights with more repetitions is used, and when size (hypertrophy) is the goal, more weight with less repetitions are used. However, there are many different ideas available, so training withing a realistic, safe, consistent manner should form the foundations of the training and training with a goal in mind and for enjoyment allows for longer term success.

Most gyms have large weight machines known as pin machines or cable machines and often these can be ideal for beginners as it allows a person to learn the basics whilst lessening the risk of injury. Most large gym equipment also contains stickers on them explaining what the machine can be used for, and some contain pictures of what muscles are used.

List of Gym Equipment Exercises

Agility Ladder Grapevine

Agility Ladder Sprint

Back Extension Hold Variation Weighted

Balance Board Burpees (Hands on board)

Balance Board Mountain Climbers

Balance Board Plank

Balance Board Plank Jacks

Balance Board Pushup – Start from a plank position with hands either side of the board. Then slowly yourself until either stomach or chest touches board. Then slowly raise self-back into plank position.

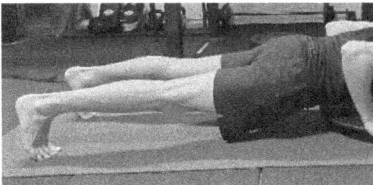

Balance Board Single Leg Squat

Balance Board Squat

Barbell Bench Press

Barbell Bench Press Variation Dead-stop

Barbell Bent Over Row

Barbell Bent Over Row Variation Single Arm

Barbell Bent Over Row Variation Snatch Grip

Barbell Bent Over Row Variation Tripod Row

Barbell Bent Over Row Variation Yates Row

Barbell Bicep Curl

Barbell Cuban Press

Barbell Cuban Rotation

Barbell Deadlift – There are several variations in Deadlift form, but this will focus on the basics or beginner interpretation. Start with either the bar only or a light weight either side. Gently squeeze shoulder blades and activate rest of core. As you lift the bar, keep core activated and press through the heels. Once at the top, slowly lower the bar until approximately shin height, maintaining strong posture throughout. Then, as you lift the bar back to standing, keep core activated once again and press through heels. During the movement keep a slight bend in the knees also known as a soft joint, so the hamstrings can still be activated, whilst minimising knee hyperextension.

Barbell Deadlift Variation Handle Addition

Barbell Deadlift Variation Single Leg

Barbell Deadlift Variation Wide/Snatch Grip

Barbell Front Squat

Barbell Hip Thrusts

Barbell Lunges

Barbell Overhead Squat

Barbell Rear Squat

Barbell Rollouts

Barbell Shoulder Press Standing

Barbell Skull Crushers

Barbell Squat Variation Hack Squat

Barbell Squat Variation Handle Hack Squat

Barbell T-Bar Row

Barbell Tripod Row

Bench Jump-Overs Hands on Bench

Bench Piriformis Stretch Standing

Bench Pullover Supine Plate Weight

Bench Triceps Dip

Box Jump-Overs

Box Jumps – Start with feet approximately hips width apart. Then use the momentum of your arms to assist in jumping higher. As you land, try to minimise the sound by activating your glute and leg muscles and slowly land into a squat position.

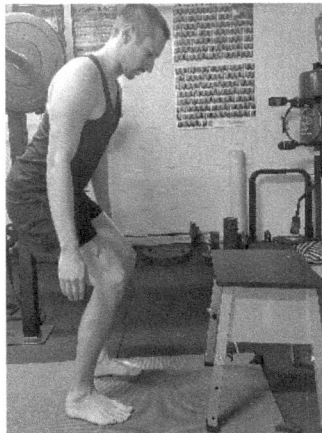

Box Jumps Variation Side Jumps

Box Step-Ups

Butterfly Stretch Seated Variation Loaded/with Weights

Cable Machine Chest Fly

Cable Machine Chest Press

Cable Machine Delt Fly

Cable Machine External Rotation

Cable Machine Face Pulls

Cable Machine Internal Rotation

Cable Machine Leg Curl Prone

Cable Machine Leg Curls Seated

Cable Machine Leg Press

Cable Machine Leg Raises

Cable Machine Palloff Press

Cable Machine Seated Lat Pulldowns

Cable Machine Seated Lat Pulldowns Variation Close Grip

Cable Machine Seated Lat Pulldowns Variation Underhand Grip

Cable Machine Seated Rows

Cable Machine Seated Rows Variation Close Grip

Cable Machine Shoulder Press

Cable Machine Single Arm Lat Pulldown

Cable Machine Single Arm Row

Cable Machine Triceps Overhead

Cable Machine Triceps Pushdown

Dumbbell Bench Chest Press

Dumbbell Bench Chest Press Variation Deadstop

Dumbbell Bench Chest Press Variation One Arm One Weight – Feet either on ground or on support bar to assist in neutral spine. Start with arm almost straight (slight bend in elbow/soft joint). Slowly lower weight until either elbow or weight is in-line with the bench, dependent on strength level, weight, and training goal. Slowly raise again to starting position and repeat until repetitions are complete, then change sides.

Dumbbell Bench Chest Press Variation One Arm Two Weights

Dumbbell Bench Pullover Supine

Dumbbell Bent Over Row Both Hands

Dumbbell Bent Over Row Variation Both Hands Reverse Grip

Dumbbell Bent Over Row Variation Single Arm

Dumbbell Bent Over Row Variation Single Arm with Bench

Dumbbell Bicep Curl

Dumbbell Bicep Curl Variation Hammer

Dumbbell Chest Fly

Dumbbell Cuban Press

Dumbbell Cuban Rotation

Dumbbell Decline Bench Chest Press

Dumbbell External Rotation

Dumbbell Flys Standing

Dumbbell Flys Standing Bent-Over

Dumbbell Front Raises Seated

Dumbbell Front Raises Standing

Dumbbell Front Raises Standing Bent-Over

Dumbbell Incline Bench Chest Press

Dumbbell Incline Bench Y-T-W

Dumbbell Incline Prone Cuban Press into Y

Dumbbell Internal Rotation

Dumbbell Leaning Lat Pulses

Dumbbell Leaning Lat Raise Hold

Dumbbell Lunges

Dumbbell Rear Delt Fly

Dumbbell Seated Forearm Curl

Dumbbell Shoulder Press Seated

Dumbbell Shoulder Press Standing

Dumbbell Shoulder Press Variation Single Arm Seated

Dumbbell Shoulder Press Variation Single Arm Standing

Dumbbell Side Raises Standing

Dumbbell Side Raises Standing Bent-Over

Dumbbell Skull Crushers

Dumbbell Triceps Kickbacks

Dumbbell Triceps Overhead

Dumbbell Tripod Row

Elliptical Trainer

Fit Ball Decline Plank Forearms

Fit Ball Decline Plank Hands

Fit Ball Incline Plank Forearms

Fit Ball Incline Plank Hands

Fit Ball Leaning Plank Bounce

Fit Ball Leaning Shin Swim

Fit Ball Prone Knees to Chest

Fit Ball Seat

Fit Ball Seat Variation Balance, Feet off Ground

Fit Ball Seiza (shins on Fit Ball) – Shins on ball, glutes on heels, activate your core, find your balance point, and hold position.

Fit Ball Sit-ups

Fit Ball Toes to Hands Throw Sit-ups

Fit Ball V-Sit-ups

Foam Roller Calf Release

Foam Roller Core Activation

Foam Roller Forearm Release

Foam Roller Gluteal Release

Foam Roller Hamstring Release

Foam Roller Hip Flexor Release

Foam Roller Hip Flexor Release Bent Leg

Foam Roller ITB Release

Foam Roller Knee Raise

Foam Roller Knee Raise Arms to Ceiling Hold

Foam Roller Overhead Arms Supine

Foam Roller Piriformis Release

Foam Roller Quadricep Release

Foam Roller Thoracic Extension

Foam Roller Tibialis Release

Goblet Squat

Good Mornings

Heavy Rope Climb Double/Unilateral Rope Climb

Heavy Rope Climb Double/Unilateral Rope Variation L-Sit

Heavy Rope Climb Single Rope

Heavy Rope Climb Single Rope Variation L-Sit

Hexagon Agility Test

Indoor Bike

Kettle Bell Standing Overhead Hold

Kettle Bell Standing Overhead Shoulder Press

Kettle Bell Standing Overhead Walking

Kettle Bell Turkish Swing

Kettle Bell Walking Forearm Extensor Pulses – Hold kettle Bell in fingertips, then slowly raise fingers into a fist like shape, then slowly extend finger.

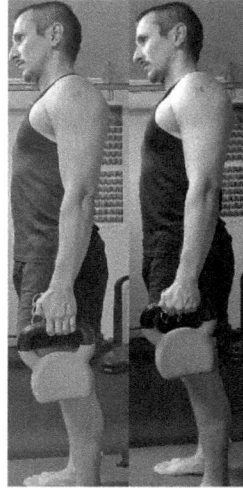

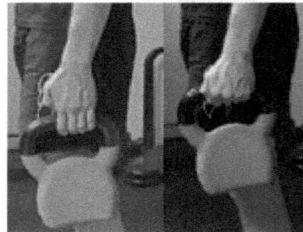

Kettlebell Farmer's Walk / Suitcase Walk

Leg Press Machine 45 degrees

Machine Preacher Curl

Seated Good Mornings

Body Weight, Calisthenics & Gymnastic Rings

Programming tips for writing Calisthenics programs will be using a log of information from the AMN academy holistic health course. The first idea that should be incorporated in calisthenic related training is incorporating locked joints in isometric holds. Traditionally the fitness industry has taught us that locking joints is wrong and dangerous. However, it is now believed that locking joints is only dangerous when lifting heavier weights. When practicing body weight movements, it is actually recommended to lock joints as long as the person doesn't have any pre-existing injuries. This locking of the joints will actually increase strength output via strengthening the connective tissue.

Correct or normal range of motion is the healthy range of motion completed during a body weight exercise without negatively impacting posture. For example, a handstand should be done with 180 degrees range of motion with a healthy strong posture. Flexibility also allows for healthier range of motion and easier strength output.

The last relevant aspect is body awareness. Body awareness allows for healthier movement, greater muscle activation and muscle control, and better technique. This can improve with body and brain related training as well as longer-term practice of body weight exercises.

When programming calisthenics and gymnastic rings the definition of core strength may be different to the way it is traditionally referred to in gym related fitness circles. Core strength in calisthenics can include the legs, all the way through to the rhomboids and latissimus dorsi muscles. This should be remembered, as doing a day focusing on only the abdominal muscles may be contraindicated. Instead focusing on specific holds, movement patterns or whole-body related training is recommended.

List of Body Weight, Calisthenics & Gymnastic Rings Exercises

Abdominal Wheel Rollouts Full

Abdominal Wheel Rollouts Kneeling

Abdominal Wheel Unilateral Rollouts

Abdominal Wheel Unilateral Rollouts Kneeling – Start in a kneeling plank-like position with hands on rollers. Then slowly lower the body down as the arms straighten away from the body. Lower self as low as safe, then slowly pull body back into starting position.

AMN Academy Barrel Rolls

AMN Academy Hamstring Stretch Variation Foot Angled and Extended

AMN Academy Hamstring Stretch Variation Foot Flexed

AMN Academy Resistance Band Figure 8's

AMN Academy Resistance Band Load Stretch

AMN Academy Spiral Stretch

Ankle Circles

Arm Circles with Bar/Pole Standing

Arm Circles with Bar/Pole Standing Bent Over

Arm Circles with Resistance Band Standing / Shoulder Dislocations

Arm Circles with Resistance Band Standing Bent Over

Arm Circles with Thumb Rotation

Arm Side Raises

Back Extension Variation Mission Impossible

Back Extension Variation Superman

Back Leg Scale

Back Lever

Back Lever Variation Advanced Tuck

Back Lever Variation Legs Bent

Back Lever Variation Legs Bent Straddle

Back Lever Variation Straight Legs Straddle

Back Lever Variation Tuck

Backward Shoulder Roll

Bear Crawl

Burpees

Burpees Variation Advanced Military / Pushup and Tuck Jump

Burpees Variation Beginner Military / Pushup

Burpees Variation Burpee and Squat

Burpees Variation Devil Maker

Burpees Variation Hip Thrust

Burpees Variation Sprawl / Wide Base

Calf Raise Standing Variation Tibia Raise

Calf Raises Standing

Candlestick

Candlestick Weight Assisted

Cardio Weighted Boxing Cross

Cardio Weighted Boxing Hook

Cardio Weighted Boxing Jab

Cardio Weighted Boxing Uppercut

Chest Dip

Chicken Wing Seated Stretch

Child's Pose Variation Incline Bench Neutral Grip

Chin Tuck

Chin-up – Start with palms facing towards face. Slowly raise the body until either your chin or chest reach the bar. Then slowly lower the body to start position.

Chin-up Variation Clap Above Bar

Chin-up Variation Concentric

Chin-up Variation Eccentric

Chin-up Variation Eccentric Weighted

Chin-up Variation Headbangers

Chin-up Variation L-Sit

Chin-up Variation Typewriter

Chin-up Variation Weighted

Circus Aerial Human Flag – Similar to the Human flag, the top hand will be pulling as the bottom arm pushes. Since it is on a moving pole, shoulder and core control will be more difficult.

Circus Aerial Trapeze Bar Hang / Bat Hang – The bar will be behind the knees, and hamstrings will be flexed.

Circus Aerial Trapeze Bar Sit-ups / Bat Hang Sit-ups – From the Bat Hang position, the core will be used to pull the body towards the bar as the hands reach for the rope. Once the rope is held in the hands, slowly lower self to continue repetitions. To finish, the person will be seated on top of the trapeze bar.

Crab Walk

Cross Arm Stretch

Cross Over Crunches

Crunches

Dead Hang – Set up as if a pull-up was going to be completed. Slowly lift legs off ground as shoulder blades are squeezed, and tense/activate the leg muscles. Hold position for as long as is safe, often up to a minute. Then slowly lower the body so feet are comfortably on ground.

Dead Hang Variation False Grip

Dead Hang Variation Knees Lift and Hold

Dead Hang Variation Knees to Chest

Dead Hang Variation L-Sit

Dead Hang Variation Scapula Squeeze

Dead Hang Variation Scapular Shoulder Circles

Dead Hang Variation Single Arm

Dead Hang Variation Straight Leg Lift

Dead Hang Variation Straight Leg Lift Toes to Bar

Dead Hang Variation Under Hand Grip

Dead Hang Variation Windscreen Wipers

Deadman Squats

Decline Plank Medicine Ball

Decline Plank Wall

Decline Plank Weight Stack or Step

Dish Hold

Dish Rocks

Dragon Flag

Dragon Flag Negatives

External Rotations Band

Figure 8 Head Circles Standing

Finger Hand Extensor Pulses

Forearm Stretch

Forearm Stretch Reverse

Forearm Stretch Variation Palm Closed Fingers Down

Forearm Stretch Variation Palm Open Fingers Down

Forward Shoulder Roll

Frog Jumps

Frog Stand

Frog Stand Variation High Frogger

Frog Stretch Hips Flat

Frog Stretch Hold

Frog Stretch Rocking

Frog Stretch Side Stretch

Front Leg Scale

Front Lever

Front Lever Variation
Advanced Single Leg Lever

Front Lever Variation
Advanced Tuck Lever

Front Lever Variation Band
Assisted on Hips

Front Lever Variation One Leg
Bent

Front Lever Variation Single
Leg Lever

Front Lever Variation
Tabletop Lever

Front Lever Variation Tuck
Lever

Glute Hamstring Martin St
Louise Nordic Lifts

Glute Hamstring Raises

Gymnastic Ring Static Hold
Variation Arms away from
Body Palms Facing Back
(Moderate)

Gymnastic Ring Static Hold
Variation Bicep Hold
(Moderate)

Gymnastic Ring Static Hold
Variation Palms Facing Body
(Easiest)

Gymnastic Rings Back Lever

Gymnastic Rings Basket
Stretch Hold

Gymnastic Rings Chest Dip

Gymnastic Rings Front Lever

Gymnastic Rings German Hang – Often used as a warmup or progression on the way to the 'Skin the cat' exercise. Begin as if a 'Skin the cat' was going to be completed. Stop when body has rotated through the gap between the arms. Hold position for as long as safe, then if safe to do so, slowly release grip.

Gymnastic Rings Inverted Deadlifts

Gymnastic Rings Inverted Hold

Gymnastic Rings Inverted Pullups

Gymnastic Rings Inverted Splits

Gymnastic Rings Knees Hold

Gymnastic Rings L-Sit Hold

Gymnastic Rings Plank

Gymnastic Rings Pullups Chinup Grip

Gymnastic Rings Pullups Neutral Grip

Gymnastic Rings Pullups Wide Grip

Gymnastic Rings Pushup

Gymnastic Rings Pushup Decline

Gymnastic Rings Pushup Arrows

Gymnastic Rings Scarecrow

Gymnastic Rings Skin the Cat Bent Legs

Gymnastic Rings Skin the Cat Straight Leg

Gymnastic Rings Supinated Pullups

Gymnastic Rings Supine Rows

Gymnastic Rings Upside Hold
– Start sitting under the rings. Then slowly lift and rotate body until the body is upside down. Trying to visualise/hold a straight line from toes through pelvis to shoulders. Head can be looking forward or slightly looking towards toes (as in picture).

Gymnastic Rings Y-T-W

Hamstring Static Stretch Variation Hips Closed (Slight Twist)

Hamstring Static Stretch Variation Hips Square/Inline

Hamstring Static Stretch Variation Open Hips

Hamstring Supine Figure 8 Stretches Band/Towel

Hand Grip Trainer Squeeze

Handstand Hold Variation Offset

Handstand Hold Variation Single Arm Wall Assisted

Handstand Hold Wall Assisted

Handstand Progression AMN Academy

Handstand Progression Shoulder Width Stance Bent Arm Lift

Handstand Progression Shoulder Width Stance Straight Arm Lift

Handstand Progression Wide Stance Bent Arm Lift

Handstand Progression Wide Stance Straight Arm Lift

Handstand Pushups Wall Assisted

Handstand Variation Offset
Wall Pushups

Hanging Hip Hiker – Begin
from normal pullup position.
Often this will be with hands
slightly wider than shoulder
width. Then slowly release on
hand and allow legs to move
to side. Hold position.

Hanging Leg Raises Variation Legs Bent

Hanging Leg Raises Variation Straight Leg

Hanging L-Sit Hold

Hanging Pike Stretch

Hanging Wind Screen Wipers

Headstand Straight Legs

Headstand Variation Forearm Base

Headstand Variation Knees Bent

Headstand Variation Single Leg Straight

Headstand Variation Tripod Base

High Knees

Hindu Squat

Hip Hurdles

Hip Hurdles Reverse

Hollow Body Hold

Hollow Body Rocks

Hops

Human Flag Chamber Hold Regression

Human Flag Double Bent Knees

Human Flag Hug Regression Hug Pole – Top arm will wrap around pole. Bottom arm will be bent and have side of body resting on triceps. Both hands will be doing a variation of a pistol grip.

Human Flag Single Bent Knee

Human Flag Support Press Regression

Human Flag Vertical Hold Regression

Human Flag – Gently squeeze shoulder blades, top hand will pull body weight as bottom hand pushes into pole. Once in position the whole body from the neck to the toes should be activated to hold one solid position. Bottom hand will normally be a pistol grip.

Ice Cream Makers

Incline Bench DB "Y" Hold Stretch

Incline Mountain Climbers

Incline Plank

Incline Plank Medicine Ball

Incline Plank Weight Stack or Step

Internal Rotations Band

Inverted Deadlifts on Bar

Jogging

Knee Circles Standing Feet Together

Knee Circles Standing Single Leg

Lateral Bounds

Lateral Hops

Latissimus Dorsi Static Stretch

Low Arm Pectoralis Major Stretch Standing

Lunge Stretch Variation Pelvic Lift with Weight Above Head

Lunge Stretch Variation Pelvic Lift with Weight Above Head Pulsing Forward

Lunge Stretch Variation Thread the Needle

Lunges

Lunges Plyometric Variation 180 Rotation

Lunges Plyometric Variation Leg Stay Same

Lunges Plyometric Variation Legs Alternate

Lunges Variation Incline Front Leg

Lunges Variation Petersen Inspired Lunge, also known as ATG Lunge

Lunges Variation Walking

Medicine Ball Slams

Medicine Ball Walking Lunges

Medicine Ball Walking Lunges Variation Overhead

Mountain Climbers

Muscle Up Negatives

Muscle Ups

Neck Soft Nods/Deep Neck Flexor Stretches Supine

NFL Inspired Agility Test

Open-Close Supine

Pendulum Supine

Planche Regression ½ Legs

Planche Regression Bent Arm – Start by having elbows in-line with edge of midsection. Slowly lift legs simultaneously off ground. Midsection is gently resting on triceps and elbows.

Planche Variation Band Assisted

Planche Variation Box Hold

Planche Variation Frog

Planche Variation Pushup

Planche Variation Straddle

Planche Variation Tuck

Planche Variation Tuck Swing

Plank Forearms

Plank Hands

Plank Variation Alternating Knees to Ribs

Plank Variation Hip Tap

Plank Variation Jacks

Plank Variation Kick

Plank Variation Romper Stompers

Plank Variation Shoulder Tap

Plank Variation Towel Slides

Plank Variation Wrist Tap

Plank Walk

Plank Walk Variation on Hand Blades

Plate Weight Overhead Abdominal Twist Seated

Plate Weight Standing Truck Driver

Plate Weight Steering Wheel Variation Bent Over

Plate Weight Steering Wheel Variation Front Raise Position

Plate Weight Steering Wheel Variation Overhead

Prone Quad Stretch Single Leg

Prone T

Prone T Variation Standing

Prone W

Prone W Variation Standing

Prone Y

Prone Y Variation Standing

Pullup Variation ½ Pullup/Inverted Rows

Pullup Variation 3 Step Pause

Pullup Variation 90 Degree Hold – Set up for the normal pullup position. Often this will be with hands slightly wider than shoulder width. Slowly pullup body upwards until the forearms and elbow crease make an approximately 90-degree angle. Hold position for as long as is safe, then slowly lower body down to starting position.

Pullup Variation Archer

Pullup Variation Clap Above Bar

Pullup Variation Commander Close Grip

Pullup Variation Commander Shoulder Width Grip

Pullup Variation Concentric

Pullup Variation Eccentric

Pullup Variation Eccentric Weighted

Pullup Variation False Grip

Pullup Variation Headbangers

Pullup Variation L-Posture

Pullup Variation L-Sit

Pullup Variation Over-Under

Pullup Variation Prison Style

Pullup Variation Single Arm

Pullup Variation Single Arm – Band Assisted with one arm

Pullup Variation Single Arm Arm-Assisted Chinup Grip

Pullup Variation Single Arm Hold– Band Assisted with one arm

Pullup Variation Switch Grip

Pullup Variation Thumbless Grip

Pullup Variation Tuck Front Lever

Pullup Variation Typewriter

Pullup Variation Weighted

Pullup Variation Weighted Chinups

Pullup Variation with Towel – Start with a strong towel over the bar. Grip the towel in each hand and slowly raise body upwards until face is in-line with hands. Briefly hold, then slowly lower body to start position.

Pullups

Pushup On Knees

Pushup Row Dumbbell

Pushup Variation 'T' / Arrow Medicine Ball

Pushup Variation 'Y' Medicine Ball

Pushup Variation Bicep – Start with fingers pointing towards toes. If possible, begin with wrists in line with shoulders. Slowly lower chest towards ground as low as is safe. Then slowly raise back to start position.

Pushup Variation Chest Tap

Pushup Variation Clap

Pushup Variation Diamond Pushup

Pushup Variation Eccentric Phase

Pushup Variation Gecko

Pushup Variation Pike/Shoulder Pushup

Pushup Variation Pseudo Planche Pushup

Pushup Variation Pseudo Superman

Pushup Variation Semi-Circles

Pushup Variation Single Arm

Pushup Variation Single Arm

Pushup Variation Single Arm Eccentric, Both Arms Concentric

Pushup Variation Sky Diver Hold

Pushup Variation Sky Diver Slide

Pushup Variation Spiderman

Pushup Variation Tiger Bend – Here I have done the easier variation for starting. Hands are in-line with face, and elbows are pulled in. While keeping hands and wrists on ground, slowly lift body until in pull hands plank position. Then slowly lower back to start. To increase difficulty, start with hands further away from face.

Pushup Variation Tiger Bend Incline

Pushup Variation Tiger Bend on Knees

Pushup Variation Triceps

Pushup Variation Wide Grip

Pushups

Pushups Variation Offset

Pushups Variation Parallette Depth

Quadricep Flex

Quadricep Flex Seated Variation Leg Raise

Quadricep Stretch Hold

Quadratus Lumborum Stretch Hold Variation Feet parallel and angled Left

Quadratus Lumborum Stretch Hold Variation Feet parallel and angled Right

Quadratus Lumborum Stretch Hold Variation Feet parallel and Straight

Quadrupe Hold

Quadrupe Kick Through

Reclining Twist

Reclining Twist Weighted

Resistance Band Chest Pull-apart

Resistance Band Overhead Pull-apart

Resistance Band Standing Delt Fly

Resistance Band Triceps Pushdown

Rhomboid Pushups

Russian Twist

SCM Neck Stretch

Seated Abdominal Twist

Seated Half Frog Stretch Hold

Seated Leg Raises

Seated Pike Hold

Seated Pike Lifts

Seated Pike Pulses

Seated Pike Stretch Nose to Shin

Seated Straddle Lifts

Seated Straddle Variation Extensions

Seiza Hold

Seiza to Lunge to Standing

Shoulder Circles with Arm Straight (Like a Train)

Side Bends

Side Bends Variation Leg Crossed in Front

Side Stepping

Side/Oblique Plank

Side/Oblique Plank Variation Forearm

Side/Oblique Plank Variation Knee to Elbow

Side/Oblique Plank Variation Star Hold

Side/Oblique Plank Variation Star Pulse

Side/Oblique Plank Variation

Side-lying Shoulder External Rotation Variation with Dumbbell

Single Bar Chest Dip

Single Bar Chest Dip Hold

Single Bar Chest Dip Variation
Korean Dip

Single Bar Chest Dip Variation
Underhand Grip (chinup grip)

Single Bar Chest Dip Variation
Underhand Grip (chinup grip)

Single Leg Calf Raise Standing

Single Leg Calf Raise Standing
Variation Tibia Raise

Single Leg Squat

Single Leg Squat Variation
Band Assisted

Single Leg Squat Variation
Dragon Pistol Squat

Single Leg Squat Variation
Dragon Pistol Squat Assisted

Single Leg Squat Variation
Pistol Squat

**Single Leg Squat Variation
Pistol Squat Assisted** – The
resistance band is to assist in
form and strength building
until the person is ready to
attempt the exercise without
the assistance. Slowly lower
the body as one leg slowly
moves in front of the body.
Finish once the glutes are as
close to the squatted heel as
possible. Slowly raise back to
start, ensuring the resistance
bands are used secondary as
needed. The amount of arm
activation will depend on how
much assistance the leg
needs in lowering and raising
safely.

Sit-up

Sit-up Variation Bicycle

Skipping with Rope

Slackline Walking

Sleeper Stretch

Sliding Wall Squat

Speed Punching Straight Punch

Sprinting

Squat Jacks

Squat Pulses

Squat Pulses/Slides Wall Assisted

Squat Seat

Squat Seat Variation Calm Raise

Squat Seat Variation Wall Assisted

Squat Seat Variation Wall Assisted & Weighted

Squat Variation Kicks

Squat Variation Shrimp Squat Elevated

Squat Variation Side Lunge Squat

Squat Variation Sissy Squat

Squats

Squats Variation Close Squats

Standing Lever Over ITB Stretch

Standing Plyometric Hamstring/Leg Swing Variation Straight Leg up and down

Standing Plyometric Hamstring/Leg Swing Variation Straight Leg up and Bent Leg down

Standing Plyometric Side Leg Swing

Standing Shoulder External Rotation 90 degrees Abduction Variation with Dumbbell

Star Jumps

Tai Chi Ankle Warm Up Stretches

Thoracic Mobilisation Supine Horizontal Variation with Arm Stretch

Toe Lifts 4 Small Toes

Toe Lifts Big Toe

Toe Side Lift 4 Small Toes

Toe Side Lift Big Toe

Toe Touch Standing

Toe Touch Standing Variation One Leg in Front

Torso Twist

Trap Stretch Standing

Tuck Jumps

V-Sit-up – Start with hands and feet straight and core activated. Slowly lift upper and lower body towards the top at a similar rate. Then slowly lower back to start position.

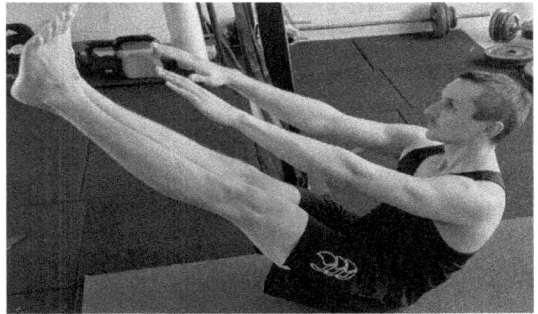

V-Sit-up Hold

V-Sit-up Variation (Head-Shoulders-Knees)

V-Sit-up Variation Side V-Ups

V-Sit-up Variation Single Leg

V-Sit-up Variation Straddle

V-Sit-up Variation X-Up

Waiter Stretch Standing

Weighted Bench Pull Overs

Weighted Plank

Weighted Pullups

Wide Squat Seat

Wide Squat Seat Variation with Calm Raise

Wide Squat Towel Slides

Wood Chops Standing

Wrist Warmup Variation Back of Hand on Ground Fingers Backward – Begin in a four-point position. Then slowly change hands so that the top of the hands is on the ground and the fingers are pointing towards the knees. This can be done as a hold for a static stretch, or as gentle pulsing forward and backward for a mobility type stretch.

Wrist Warmup Variation Back of Hand on Ground Fingers Backward, Sidewards Pulsing

Wrist Warmup Variation Fingers Forward, Forward Pulsing

Wrist Warmup Variation Palms on Ground Fingers Backward, Forward Pulsing

Wrist Warmup Variation
Palms on Ground Fingers
Backward, Sidewards Pulsing

Wrist Warmup Variation
Palms on Ground Fingers
Forward, Sideward Pulsing

Wrist Warmup Variation
Palms on Ground Fingers
Sideward away from Body,
Forward Pulsing

Wrist Warmup Variation
Palms on Ground Fingers
Sideward away from Body,
Sidewards Pulsing

Yoga

As I am not a yoga instructor, only a yoga student, I will not include programming tips in this section. There are various types of yoga practice, different goals, etc., and I believe a student of yoga suggesting programming tips would not only be ethically incorrect but may be dangerous.

List of Yoga and Yoga Inspired Exercises

2 Prasarita Padottanasana Variations

3 Pigeon Pose Stretch Variations

5 Star Squat Malasana

Bhujapidasana

Bhujapidasana Shoulder Prep

Bow Quad Stretch Standing

Buddha Prayer Deep Squat – Start in a low wide squat. Gentle squeeze shoulder blades as the elbows gently push into the inside thighs. The hands will gently touch forming the prayer-like position with the hands. Hold the position.

Buddha Squat Bows

Child's Pose Variation (Arms by side)

Crow Pose Bakasana

Downward-Facing Dog

Eagle Garudasana

Elevate Leg Position Legs Together

Elevate Leg Positive Legs Apart Variation

Firefly Prep Standing

Firefly Tittibhasana Resistance Band Assisted

Five Star Squat

Half Moon Sequence

Half Moon Sequence Triangle Hold Standing

Headpull

High Lunge Hold

Knee to Ankle Pose – Agnistambhasana – Sit with left leg on ground making an approximately 60-degree bend between the knee and the hamstring. Then place the right leg on top of the left leg. Hold position, until time to swap legs.

Knee to Ear / Karnapidasana

Lizard Pose Utthan Pristhasana

Low Lunge Stretch Variation

Plow Pose / Halasana

Separate Leg

Separate Leg Head to Knee

Single Leg Straight Leg Hold
(Toes in Fingers)

Supta Padangusthasana

Supta Padangusthasana
Variation Leg Across the Body

Supta Padangusthasana
Variation Leg Open

Tiger Pushup

Triangle

Upavista Konasana

Upward-Facing Dog/Cobra

Utthita Hasta
Padangustahasana Standing

<u>Appendix/References</u>

Formal References

Abrahamson, E., & Langston, J. (2017). Making Sense of Human Anatomy and Physiology A Learner-Friendly Approach.

Academis, N. (2005). Dietary reference intakes for energy, carbohydrates, fiber, fat, fatty acids, cholesterol, protein, and amino acids., *Vol. 5.*, National Academy Press

Acid, P. Fact Sheet for Health Professionals". National Institutes of Health [Internet]. Available: https://ods.od.nih.gov/factsheets/PantothenicAcid, HealthProfessional.

Adams, A. (1973). Ninja: The Invisible Assassins. Ohara Publications Inc.

Aisbett, B. (2013). Fixing It: The Complete Survivor's Guide to Anxiety-Free Living. Harper Collins Inc.

Akinrodyoye, M. A., & Lui, F. (2020). Neuroanatomy, Somatic Nervous System

Albert, P. R. (2010). Epigenetics in mental illness: Hope or hype?. *Journal of psychiatry & neuroscience: JPN, 35*(6), 366.

Alexander, P. (1994). It Could Be Allergy and It Can Be Cured

Arden, J. B. (2010). Rewired Your Brain: Think Your Way to a Better Life. Wiley.

Arnow, L. E. (1976). Introduction to Physiological and Pathological Chemistry. Mosby Inc.

Bailie, J. M. (1984). Giant Book of Knowledge. Octopus Publishing Group.

Barret, K. E. (2010). Ganong; s Review of Medical Physiology. USA.

Becker, R. O., Selden, G. (1998). The Body Electric: Electromagnetism and The Foundation of life

Benowicz, R. J. (1983). Vitamins & You. Grosset & Dunlap.

Bodri, B., & Newtson, J. (2011) Internal Martial Arts Nei-gong Cultivating Your Inner Energy to Raise Your Martial Arts to the Next Level. Top Shape Publishing.

Bohlander, A., Geweniger, V. (2014). Pilates – A Teacher's Manual: Exercises with Mats and Equipment for Prevention and Rehabilitation

Borghuis, J., Hof, A. L., & Lemmink, K. A. (2008). The importance of sensory-motor control in providing core stability: implications for measurement and training. *Sports Medicine, 38, 893-916.*

Bowman, P. (2019). Deconstructing martial arts (p. 182). Cardiff University Press.

Brannon, L., & Feist, P. (1996). Health Psychology: An Introduction to Behaviour and Health. *Third Edition.* Wadsworth Publishing.

Burnham, T. (2001). Mean Genes: From Sex to Money to Food – Taming Our Primal Instincts. Simon & Schuster.

Burton, L., Westen, D., & Kowalski, R. (2006). Psychology: Australian and New Zealand Edition. John Wiley & Sons, Inc.

Burton, L., Westen, D., & Kowalski, R. (2008). Psychology 2: Australian and New Zealand Edition. John Wiley & Sons, Inc.

Carey, N. (2012). Epigenetics Revolution: How Modern Biology is Rewriting our Understanding of Genetics, Disease and Inheritance. Icon Books.

Case-Smith, J., Weaver, L. L., & Fristad, M. A. (2014). A systematic review of sensory processing interventions for children with autism spectrum disorders. *Autism: The International Journal of Research and Practice.*

Ch'oe, H. (1998). Hap Ki Do: The Korean Art of Self Defense.

Chaplan, J. P. (1970). Dictionary of Psychology. Dell Publishing.

Chaurasia, B. D. (2004). Human anatomy (p. 53). New Delhi, India: CBS Publisher.

Chia, M. (1986). Iron Shirt Chi Kung I: Once a Martial Art, Now the Practice that Strengthens the Internal Organs, Roots Oneself Solidly, and Unifies Physical, Mental, and Spiritual Health. Universal Tao Publications.

Cho, S. H. (1969). Self-Defense Karate. Stravon Educational Press.

Chutkan, R. (2016). Microbiome Solution: A radical new way to heal your body from the inside out. Scribe Publications.

Clark, R. (2012). Pressure-point Fighting: A Guide to the Secret Heart of Asian Martial Arts. Tuttle Publishing.

Cohen, M., J., Schreiner, R. (2007). Reconnecting with Nature: Finding wellness through restoring your bond with the Earth.

Cooper, E. L. (2003). Neuroimmunology of autism: a multifaceted hypothesis. *International journal of immunopathology and pharmacology, 16*(3), 289-292.

Crudelli, C. (2008). *The Way of the Warrior.* Dorling Kindersley Ltd.

Cummins, A. (2016). Samurai and Ninja: The Real Story Behind the Japanese Warrior Myth that Shatters the Bushido Mystique. Tuttle Publishing.

D'Souza, C., (n.d.)., Good Health Through Spices, Herbs and Indian Dishes

Davis Jr, G. E., & Lowell, W. W. (2006). Solar Cycles and their relationship to human disease and adaptability. *Medical hypotheses, (67(3),* 447-461.

Doidge, N. (2008). The Brain that Changes Itself: Stories of Personal Triumph from the Frontiers of Brain Science.

Dompe, C., Moncrieff, L., Matys, J., Grzech-Leśniak, K., Kocherova, I., Bryja, A., ... & Dyszkiewicz-Konwińska, M. (2020). Photobiomodulation—underlying mechanism and clinical applications. *Journal of clinical medicine, 9*(6), 1724.

Easley, T., & Horne, S. (2016). The Modern Herbal Dispensatory: A Medicine-Making Guide

Feldenkrais, M. (1985). The potent self: A guide to spontaneity. Harper & Row.

Feldenkrais, M. (1987). Awareness Through Movement: Health Exercises for Personal Growth. Penguin Handbooks

Feldenkrais, M. (2011). Embodied wisdom: The collected papers of Moshé Feldenkrais. North Atlantic Books.

Fortuna, L. R., & Vallejo, Z. (2015). Treating co-occurring adolescent PTSD and addiction: Mindfulness-based cognitive therapy for adolescents with trauma and substance-abuse disorders. New Harbinger Publications.

Fourlanos, S., Dotta, F., Greenbaum, C. J., Palmer, J. P., Rolandsson, O., Colman, P. G., & Harrison, L. C. (2005). Latent autoimmune diabetes in adults (LADA) should be less latent., *Diabetologia, 48*, 2206-2212.

Freeberg, L. (2009). Discovering Biological Psychology. Cengage Learning.

Garg, G., Bhati, S., & Kataria, S. The role of wild plants and herbs in restoring holistic health and fighting the infections borne by the epidemic COVID-19.

Garrett, R. H., & Grisham, C. M. (2016). Biochemistry. Cengage Learning.

Goldstein, E. B. (2014). Cognitive psychology: Connecting mind, research and everyday experience. Cengage Learning.

Goleman, D. (1996). Emotional Intelligence: Why it can matter more than IQ. BloomsburyPublishing.

Goodman, F. (2015). Karate, Aikido, Ju-jitso & Judo. Anness Publishing.

Green, T. A. (2010). Martial arts of the world: an encyclopedia of history and innovation (Vol. 1). Abc-Clio.

Green, T. A. (2010). Martial arts of the world: an encyclopedia of history and innovation (Vol. 2). Abc-Clio.

Hall, J. E., & Hall, M. E. (2020). Guyton and Hall textbook of medical physiology e-Book. Elsevier Health Sciences.

Hamasaki, H. (2020). Effects of diaphragmatic breathing on health: a narrative review. *Medicines*, *7*(10), 65.

Hamilton, H., & Rose, M. B., (1984). Cardiovascular Disorders. Lippincott Williams & Wilkins.

Harbottle, L., & Schonfelder, N. (2008). Nutrition and Depression: A review of the evidence. *Journal of Mental Health, 17*(6), 576-587.

Hari, J. (2016). Chasing the Scream: The First and Last Days of the War on Drugs. Bloomsbury Publishing.

Ho'o, M. (2004). Tai Chi Chun. Black Belt Magazine Video.

Heidenstam, D., Kramer, A., Midgley, R., Sturrock., et. al. (1984). Human Body. Galley Press.

Jacob. S. W., & Francone, C. A. (1974). Structure and Function in Man. Saunders, Ken & Georgie.

Kaminoff, L., & Matthews, A. (2021). Yoga anatomy. Human Kinetics.

Kelly, R. (2007). MD, The Human Antenna: Reading the Language of the Universe in the Songs of Our Cells.

Kendroud, S., Fitzgerald, L. A,. Murray, I., & Hanna, A. (2021). Physiology, Nociceptive pathways.

Koga, R. (2004). Practical Aiki-Do Volume 1. DVD. Black Belt
	Magazine Video.

Kouka, N. (2009). Psychiatry for Medical Students and
	Residents. USA.

Lam, P., & Miller, M. (2006). Teaching Tai Chi Effectively: Simple
	and Proven Methods to Make Tai Chi Accessible to
	Everyone

Lane, S. J., Mailloux, Z., Schoen, S., Bundy, A., May-Benson, T.
	A., Parham, L. D., ... & Schaaf, R. C. (2019). Neural
	foundations of ayres sensory integration®. *Brain
	sciences*, *9*(7), 153.

Leach, R. M., Rees, P. J., & Wilmshurst, P. (1998). Hyperbaric
	oxygen therapy. *Bmj*, *317*(7166), 1140-1143.

Lee, B., & Uyehara, M. (2007). Bruce Lee's Fighting Method:
	Self-Defense Techniques. Ohara Publications,
	Incorporated.

Lee, B., & Uyehara, M. (2007). Bruce Lee's Fighting Method:
	Skill in Techniques. Ohara Publications, Incorporated.

Lee, B., & Uyehara, M. (2007). Bruce Lee's Fighting Method:
	Advanced Techniques. Ohara Publications,
	Incorporated.

Lépine, F. (2006). Qi-Gong and Kuji-In: A Practical Guide to an
	Oriental Esoteric Experience.

Lertola, J., Park, A. (2002). Anatomy of Anxiety: What triggers it
	and how the body responds., Time Magazine

Levine, J. A. (2002). Non-exercise activity thermogenesis (NEAT). Best Practice & Research Clinical Endocrinology & Metabolism, 16(4), 679-702.

Lieberman, M., & Peet, A. (2018). Marks' Basic Medical Biochemistry: A Clinical Approach.

Link, N., Chou, L., & Kasturia, S. (2011). The anatomy of martial arts: an illustrated guide to the muscles used in key kicks, strikes & throws. (No Title).

Little, J. (2016). *The warrior within: The philosophies of Bruce Lee*. Chartwell Books.

Low, S. (2011). Overcoming Gravity; A Systemic Approach to Gymnastics and Bodyweight Strength

McCann, T. (2011). *An evaluation of the effects of a training programme in Trauma Release Exercises on quality of life* (Master's thesis, University of Cape Town).

Martin, G. N., Carlson, N. R., & Buskist, W. (2010). Psychology. *Fourth Edition*. Pearson Education.

Medina, J. (2009). Brain Rules: 12 Principles for Surviving and Thriving at Work, Home and School. Pear Press.

Melegrito, J. (2008). Philippine Fighting Arts. DVD. Black Belt Magazine Video.

Metz, A. E., Boling, D., DeVore, A., Holladay, H., Liao, J. F., & Vlutch, K. V. (2019). Dunn's model of sensory processing: an investigation of the axes of the four-quadrant model in healthy adults. *Brain sciences, 9*(2), 35.

Milner, C. E. (2008). Functional Anatomy for Sport and Exercise: Quick Reference

Mol, S. (2001). Classical fighting arts of Japan: A complete guide to Koryū Jūjutsu. Kodansha International.

Myers, T. W. (2020). Anatomy Trains: Myofascial Meridians for Manual Therapists and Movement Professionals

Nestler, E. J., Peña, C. J., Kundakovic, M., Mitchell, A., & Akbarian, S. (2016). Epigenetic basis of mental illness. *The Neuroscientist*, *22*(5), 447-463.

Neumann, K., D. (2023). Your Complete Guide to the Body Chakras., https://www.forbes.com/health/body/body-chakras-guide/

Nishioka, H., & West, J. R. (2007). The Judo Textbook: In Practical Application. Black Belt Books.

Novak, J. R., Robinson, L. P., & Korn, L. E. (2021). What MFTs should know about nutrition, psychosocial health, and collaborative care with nutrition professionals. *Journal of Marital and Family Therapy,* 00, 1-21. https://doi.org/10.1111/jmft.12540

Oceana. (2007). Human Body: A comprehensive guide to the wonders of the body. Quantum Publishing Ltd.

Odendaal, J. S. (2000). Animal-assisted therapy—magic or medicine? *Journal of psychosomatic research, 49*(4), 275-280.

Oren, G. K. (2012). Anatomy of Fitness: Yoga.

Oyama, M. (1967). Vital Karate. Japan Publications Trading Co.

Patrick, R. P., & Ames, B. N. (2015). Vitamin D and the omega-3 fatty acids control serotonin synthesis and action, part 2: Relevance for ADHD, bipolar disorder, schizophrenia, and impulsive behaviour. *The FASEB Journal, 29*(6), 2207-2222.

Peale, N. V. (1990). The Power of Positive Thinking. Ebury Press.

Perlmutter, D., & Loberg, K. (2014). Grain Brain: The Surprising Truth About Wheat, Carbs, and Sugars – Your Brain's Silent Killers. Hodder & Stoughton.

Pert, C. B. (2010). Molecules of emotion: The science behind mind-body medicine. Simon and Schuster.

Pribram, K., & Gill, M. (1976). Freud's 'Project' Re-Assessed. Hutchinson & Co. Publishers.

Ptak, C., & Petronis, A. (2022). Epigenetic approaches to psychiatric disorders. *Dialogues in clinical neuroscience.*

Purves, D., Augustine, G. J., et. al. (2004). Neuroscience: Third Edition., Sinauer Associates.

Randall, J. M., Matthews, R. T., and Stiles, M. A. (1997). Resonant frequencies of standing humans. *Ergonomics, 40(9), 879-886*

Reinisch, S., Höller, J., & Maluschka, A. (2012). The Secrets of Kyusho-Pressure Point Fighting. Meyer & Meyer Verlag.

Rim, J. B., & Sheya, J. (2005). Traditional Hapkido Volume 2. DVD. Black Belt Magazine Video.

Rosko, L., Smith, V. N., Yamazaki, R., & Huang, J. K. (2019). Oligodendrocyte bioenergetics in health and disease. The Neuroscientist, 25(4), 334-343.

Sadock, B. J., Sadock, V. A., & Ruiz, P. (2015). Kaplan & Sadock's Synopsis of Psychiatry: Behavioural Sciences/Clinical Psychiatry. *Eleventh Edition*.

Santrock, J. W. (2010). Life-Span Development. McGraw-Hill Higher Education.

Scaer, R. (2011). The Body Bears the Burden: Trauma, Dissociation, and Disease.

Schauer, E., & Elbert, T. (2010). The psychological impact of child soldiering. *Trauma rehabilitation after war and conflict: Community and individual perspectives*, 311-360.

Sertić, H., Čorak, S., & Segedi, I. (2016). APPLICABLE RESEARCH IN JUDO.

Shekelle, P. G., Cook, I. A., Miake-Lye, I. M., Booth, M. S., Beroes, J. M., & Mak, S. (2018). Benefits and harms of cranial electrical stimulation for chronic painful conditions, depression, anxiety, and insomnia: a systematic review. *Annals of internal medicine, 168*(6), 414-421.

Sinicki, A. (2022). The Protean Performance System: SuperFunctional Training 2.

Sophia. (2022). Shito Ryu: A Complete list of Shito Ryu Kata with Videos., https://www.karatephilosophy. com/category/kata/shito-ryu/.

Sperling, A. P. (1992). Psychology Made Simple. Butterworth-Heinemann Limited.

Stampi, C. (1989). Polyphasic sleep strategies improve prolonged sustained performance: a field study on 99 sailors. Work & Stress, 3(1), 41-55.

Stampi, C. (1992). Evolution, chronobiology, and functions of polyphasic and ultrashort sleep: main issues. Why we nap: evolution, chronobiology, and functions of polyphasic and ultrashort sleep, 1-20.

Sun, H. H., Meng, J., & Yan, K. (Eds.). (2020). The Book of Chinese Medicine, Volume 1: The Timeless Science of Balance and Harmony for Modern Life (Vol. 1). Cambridge Scholars Publishing.

Świątczak, B. (2019). Francisco Varela's Vision of the Immune System.

Tenger, B. (1978). Self-Defense: Nerve Centres & Pressure Points for Karate, Jujutsu, and Atemi-Waza

Thomas, G. B. B., & Thomas, M. F. S. (1967). The Biology of Man. Hulton Educational Publications.

Trumbo, P., Schlicker, S., Yates, A. A., & Poos, M. (2002). Dietary reference intakes for energy, carbohydrate, fiber, fat, fatty acids, cholesterol, protein and amino acids. (Commentary). Journal of the American dietetic association, 102(11), 1621-1631.

Van der Kolk. B. A. (2015). The Body Keeps the Score: Mind, Brain and Body in the Transformation of Trauma. Penguin Books.

Vasilev, V., Meridth, S., & Ryabko, M. (2006). Let Every Breath: Secrets of the Russian Breath Masters.

Vineyard, M. (2007). How you stand, how you move, how you live: Learning the Alexander Technique to explore your mind-body connection and achieve self-mastery. Da Capo Lifelong Books.

Walsh, W. J. (2014). Nutrient power: Heal your biochemistry and heal your brain. Simon and Schuster.

Warren, M. P. (1983). Effects of undernutrition on reproductive function in the human. Endocrine Reviews, 4(4), 363 - 377.

Waugh, A., & Grant, A. (2014). Ross & Wilson Anatomy and physiology in health and illness E-book. Elsevier Health Sciences.

Wiley, M. V. (2011). Filipino martial culture. Tuttle Publishing.

Wisneski, L., & Anderson, L. (2005). The scientific basis of integrative medicine.

Wood, S. J., Allen, N. B., & Pantelis, C. (Eds.). (2009). The neuropsychology of mental illness. Cambridge University Press.

Wynn, G. H. (2015). Complementary and alternative medicine approaches in the treatment of PTSD., *Curr Psychiatry Rep.*; 17: 600

Van Assche, M. (2012). The Thymus Gland.

Yang, C., Ji, J., Lv, Y., Li, Z., & Luo, D. (2022). Application of Piezoelectric Material and Devices in Bone Regeneration. Nanomaterials, 12(24), 4386.

Young, E. (2012). Gut Instincts: The secrets of your second brain. *New Scientist, 216 (2895), 38-42*

Zehr, E. P. (2008). Becoming Batman: The Possibility of a Superhero. The Johns Hopkins University Press.

Zhong, J. J., Timofeevich, A. (2008). Training Methods of 72 Arts of Shaolin

Zoughari, K. (2010). The Ninja: Ancient Shadow Warriors of Japan. Berkeley Books.

Course/Degree/CPD References

Bachelor of Psychology group Honours – James Cook University

Certificate 3, 4, Master Trainer – Australian Institute of Fitness

Diploma Counselling – TAFE North

Psychology CPD's used in this book:

- Arielle Schwartz – PESI Australia – Complex Trauma Treatment
- Bessel van der Kolk – PESI – Online Certificate regarding 'Rewiring the Brain: Neurofeedback
- Black Dog Institute – REACH Facilitator Training
- Blue Knot Foundation – A Three-Phased Approach – 'Working Therapeutically with Complex Trauma Clients'
- Community Training Australia – Workshop for 'Body Therapies'
- Community Training Australia – Workshop for 'Understanding Grief and Loss'
- Comorbidity Guidelines online training – Management of Co-occurring alcohol and other drug and mental health conditions in alcohol and other drug treatment settings
- GriffinOT – Sensory Processing Aware Level 1
- GriffinOT – Sensory Processing Aware Level 2
- GriffinOT – Sensory Processing Aware Level 3

- Headspace online training – Developmental Disorders in Young People
- Insight Alcohol and other drug training and workforce development Queensland – Modules 1-6
- Jennifer Sweeton – PESI Australia – PTSD Trauma treatment – EMDR, CBT and Somatic-Based Interventions
- Jon Kabat-Zinn – PESI – Online Certificate regarding 'Mindfulness, Healing and Transformation: The Pain and the Promise of Befriending the Full Catastrophe
- Leslie Korn – PESI – Online Certificate regarding 'Nutrition for Mental Health'
- Linda Curran – PESI – Online Certificate regarding 'Master Clinician Series The Adverse Childhood Experiences Study
- Mental Health First Aid Australia – Standard Mental Health First Aid Facilitator fourth edition course
- Mental Health First Aid Australia – Webcast 'MHFA Auditory Hallucination Simulation'
- Online training NCETA – Ice: Training for Frontline Workers Certificate of Completion modules 1-7
- PESI Australia – Autism and Sensory Processing Disorder
- PESI Australia – Autism Meltdowns in Children and Adolescents
- PESI Australia – High-Functioning Autism

- Stephen Porges PhD – PESI Australia – Clinical Applications of the Polyvagal Theory

Advanced Master Herbalist Diploma Online – Centre of Excellence

Yoga for Mental Health – Rewire Therapy – Student

Tai Chi for Arthritis Course

White Tiger Qi Gong Courses

- Trinity System Chinese Medicine Fundamentals
- Fascia Foundations Course
- 5 Element QiYo Course
- Qi Gong for Worry and Anxiety

Informal References

AMN Academy Holistic Health Coach Course

MacroFit Inc. app – Simon Ata program – Project Calisthenics
Level 1-3

Movement Athlete fitness app –
https://themovementathlete.com/

PNI Global Awareness – PNI course and Wellness Management
information

Various Articles by Sifu Anthony Korahais –
https://flowingzen.com/

Various Articles by Yogapedia - https://www.yogapedia.com/

Various GMB Online Articles – https://gmb.io/

Various 'Nutrition with Judy' Articles –
https://www.nutritionwithjudy.com/

Various Onnit Online Articles – https://www.onnit.com/

Peter Attia – https://peterattiamd.com/podcast/

Various Podcasts and articles by Dr. Rhonda Patrick -
www.foundmyfitness.com

AshtonFitness. (n.d.). Ashton Fitness. YouTube. Retrieved 2022
-2021, from
https://www.youtube.com/@AshtonFitness

Athleanx. (n.d.). Athlean-X. YouTube. Retrieved 2018-2023,
from https://www.youtube.com/@athleanx

Aucademy6195. (n.d.). Aucademy. YouTube. Retrieved 2022-2023, from
https://www.youtube.com/@aucademy6195

Blackbelt_magazine. (n.d.). Black Belt Magazine. YouTube. Retrieved from 2012-2023, from
https://www.youtube.com/@blackbelt_magazine

CaptainTepes. (n.d.) Mike Ciardi. YouTube. Retrieved from 2019-2020, from
https://www.youtube.com/@CaptainTepes

CKFAHQ. (n.d.). Henry Sue Circular Tong Long. YouTube. Retrieved from 2018-2021, from
https://www.youtube.com/@CKFAHQ

ClearMartialArts. (n.d.). Clear's Internal Combat Arts. YouTube. Retrieved from 2015-2021, from
https://www.youtube.com/@ClearMartialArts

DifferingMinds. (n.d.). Differing Minds. YouTube. Retrieved from 2023, from
https://www.youtube.com/@DifferingMinds

Elasticsteel. (n.d.). ElasticSteel. YouTube. Retrieved from 2018-2022, from https://www.youtube.com/@Elasticsteel

Everythingwingchun1. (n.d.). Everything Wing Chun. YouTube. Retrieved from 2013-2021, from
https://www.youtube.com/@Everythingwingchun1

Fightscience. (n.d.). Fight Science. YouTube. Retrieved from 2018-2021, from
https://www.youtube.com/@fightscience

Fitnessblender. (n.d.). FitnessBlender. YouTube. Retrieved from 2018-2021, from https://www.youtube.com/@fitnessblender

FitnessFAQs. (n.d.). FitnessFAQs. YouTube. Retrieved from 2021-2023, from https://www.youtube.com/@FitnessFAQs

FlowwithAdee. (n.d.). Adison Briana. YouTube. Retrieved 2020-2021, from https://www.youtube.com/@FlowwithAdee

FoundMyFitness. (n.d.). FoundMyFitness. YouTube. Retrieved from 2020-2023, from https://www.youtube.com/@FoundMyFitness

Frank_Medrano. (n.d.). Frank Medrano. YouTube. Retrieved from 2018-2022, from https://www.youtube.com/@frank_medrano

GinaScarangella. (n.d.). Gina Scarangella. YouTube. Retrieved from 2019-2020, from https://www.youtube.com/@GinaScarangella

GMBfit. (n.d.). GMB Fitness. YouTube. Retrieved from 2018-2023, from https://www.youtube.com/@gmbfit

HowtoADHD. (n.d.). How to ADHD. YouTube. Retrieved from 2022-2023, from https://www.youtube.com/@HowtoADHD

Iamcarlpaoli. (n.d.). Carl Paoli. YouTube. Retrieved from 2018-2023, from https://www.youtube.com/@iamcarlpaoli

IzzoWingChun. (n.d.). Izzo Wing Chun. YouTube. Retrieved from 2013-2021, from https://www.youtube.com/@IzzoWingChun

JeffNippard. (n.d.). Jeff Nippard. YouTube. Retrieved from 2021-2023, from https://www.youtube.com/@JeffNippard

KaratebyJesse. (n.d.). Jesse Enkamp. YouTube. Retrieved from 2022-2023, from https://www.youtube.com/@KARATEbyJesse

KinoYoga. (n.d.). Kino Yoga. YouTube. Retrieved from 2018-2023, from https://www.youtube.com/@KinoYoga

Kurzgesagt. (n.d.) Kurzgesagt – In a Nutshell. YouTube. Retrieved from 2015-2023, from https://www.youtube.com/@kurzgesagt

Livinleggings. (n.d.). Livingleggings. YouTube. Retrieved from 2022-2023, from https://www.youtube.com/@Livinleggings

MovementProjectPT. (n.d.). Movement Project PT. YouTube. Retrieved from 2023-2023, from https://www.youtube.com/@MovementProjectPT

Neuroscientificallychallenged. (n.d.). Neuroscientifically Challenged. YouTube. Retrieved from 2021-2022, from https://www.youtube.com/@Neuroscientificallychallenged

Nicabm. (n.d.). NICABM. YouTube. Retrieved from 2020-2023, from https://www.youtube.com/@nicabm

OfficialBarstarzz. (n.d.). OfficialBarstarzz. YouTube. Retrieved
     from 2018-2022, from
     https://www.youtube.com/@OfficialBarstarzz

OFFICIALTHENXSTUDIOS. (n.d.). THENX. YouTube. Retrieved
     from 2019-2022, from https://www.youtube.com/
     @OFFICIALTHENXSTUDIOS

Portaldo. (n.d.). Ido Portal. YouTube. Retrieved from 2020-2023,
     from https://www.youtube.com/@portaldo

PsychAlive. (n.d.). PsychAlive. YouTube. Retrieved from 2021
     -2022, from https://www.youtube.com/@PsychAlive

Psychologyofeating. (n.d.). Institute for the Psychology of
     Eating. Youtube. Retrieved from 2023, from
     https://www.youtube.com/@Psychologyofeating

RealMichaelJaiWhite. (n.d.). Real Michael Jai White. YouTube.
     Retrieved from 2019-2022, from
     https://www.youtube.com/@RealMichaelJaiWhite

Reflexionyogaonline. (n.d.). Reflexion Yoga. YouTube. Retrieved
     from 2018-2021, from
     https://www.youtube.com/@Reflexionyogaonline

SciShow. (n.d.). SciShow. YouTube. Retrieved from 2015-2023,
     from https://www.youtube.com/@SciShow

SenseiSeth. (n.d.). Sensei Seth. YouTube. Retrieved from 2021-
     2023, from https://www.youtube.com/@SenseiSeth

Sidpaulson1. (n.d.). Sid Paulson. YouTube. Retrieved from 2022-
     2023, from https://www.youtube.com/@sidpaulson1

Socialworkact8660. (n.d.). Social Work & ACT. YouTube. Retrieved from 2020-2022, from https://www.youtube.com/@socialworkact8660

Strengthcamp. (n.d.). Elliott Hulse's Strength Camp. YouTube. Retrieved from 2021-2023, from https://www.youtube.com/@strengthcamp

Strengthoversize. (n.d.). StrengthOVERsize. YouTube. Retrieved from 2020-2021, from https://www.youtube.com/@strengthoversize

Synergyfitnessteam1. (n.d.). Synergyfitnessteam. YouTube. Retrieved from 2020-2022, from https://www.youtube.com/@synergyfitnessteam1

TEDEd. (n.d.). TED-Ed. Youtube. Retrieved from 2015-2023, from https://www.youtube.com/@TEDEd

TheBioneer. (n.d.). The Bioneer. YouTube. Retrieved from 2022-2023, from https://www.youtube.com/@TheBioneer

TheKneesovertoesguy. (n.d.). The Kneesovertoesguy. YouTube. Retrieved from 2022-2023, from https://www.youtube.com/@TheKneesovertoesguy

TomMorrison. (n.d.). Tom Morrison. YouTube. Retrieved from 2022-2023, from https://www.youtube.com/@TomMorrison

Treforall312. (n.d.) TRE FOR ALL. YouTube. Retrieved from 2021-2023, from https://www.youtube.com/@treforall312

Westmoretonmartialarts. (n.d.). West Moreton Academy of Martial Arts. YouTube. Retrieved from 2018-2021,

from https://www.youtube.com/
@Westmoretonmartialarts

YiannisChristoulas. (n.d.). Yiannis Christoulas. YouTube.
Retrieved 2023.
http://www.youtube.com/@YiannisChristoulas

Yogabycandace. (n.d.). Candace Cabrera Tavino. YouTube.
Retrieved from 2018-202, from
https://www.youtube.com/@yogabycandace

YogaBody. (n.d.). YogaBody.Official. YouTube. Retrieved from
2023, from
https://www.youtube.com/
@YogaBody.Official/featured

YogaVibes. (n.d.). YogaVibes. YouTube. Retrieved from 2018-
2022, from https://www.youtube.com/@YogaVibes

ZuzkaLight. (n.d.). Zuzka Light. YouTube. Retrieved from 2018-
2022, from https://www.youtube.com/@ZuzkaLight

Previous Influential Professionals

Gym Manager – Donna Hartley

Long Term Personal Training Client – Janelle Fox

Mental Health Facilitator – Philippa Harris

Mental Health Manager – Alison Fairleigh

Mental Health Manager – Cassandra Parry

Personal Training Course Instructor – Rebecca Leddy

Personal Training Manager / Zen Do Kai Karate / BJC Muay Thai Instructor – Marco Vogel

Pilates Mentor – Kat Syzmanski

Psychology Manager – Liezel Gordon

Psychology Manager – Suzy Dormer

Psychology Supervisor – Gayle Roe

Psychology Supervisor – Kirsten Seymour

Science Teacher / Itosu Shito Ryu Karate Instructor – Murray Burrows

Wing Chun Instructor – Pablo Cardenas

Various other Athletics Volunteers, Fitness Colleagues, Psychology Colleagues, School Coaches and School Teachers

Many partners of the professionals who helped me with my knowledge journey.

www.ingramcontent.com/pod-product-compliance
Lightning Source LLC
Chambersburg PA
CBHW072054020426
42334CB00017B/1504